Managing Knowledge in the Construction Industry

Spon Research

publishes a stream of advanced books for built environment researchers and professionals from one of the world's leading publishers.

Published:

Free-standing Tension Structures: From Tensegrity Systems to Cable-Strut Systems
978-0-415-33595-9
B.B. Wang

Performance-based Optimization of Structures: Theory and Applications
978-0-415-33594-2
Q.Q. Liang

Microstructure of Smectite Clays and Engineering Performance
978-0-415-36863-6
R. Pusch and R. Yong

Procurement in the Construction Industry
978-0-415-39560-1
W. Hughes et al.

Communication in Construction Teams
978-0-415-36619-9
C. Gorse and S. Emmitt

Concurrent Engineering in Construction
978-0-415-39488-8
C. Anumba

People and Culture in Construction
978-0-415-34870-6
A. Dainty, S. Green and B. Bagilhole

Very Large Floating Structures
978-0-415-41953-6
C.M. Wang, E. Watanabe and T. Utsunomiya

Tropical Urban Heat Islands: Climate, Buildings and Greenery
978-0-415-41104-2
N.H. Wong and Y. Chen

Innovation in Small Construction Firms
978-0-415-39390-4
P. Barrett, M. Sexton and A. Lee

Construction Supply Chain Economics
978-0-415-40971-1
K. London

Forthcoming:

Location-based Management System for Construction: Improving Productivity Using Flowline
978-0-415-37050-9
R. Kenley and O. Seppanen

Employee Resourcing in the Construction Industry
978-0-415-37163-6
A. Raiden, A. Dainty and R. Neale

Managing Knowledge in the Construction Industry

Alexander Styhre

Routledge
Taylor & Francis Group

LONDON AND NEW YORK

First published 2009
by Taylor & Francis

2 Park Square, Milton Park, Abingdon, Oxon OX14 4RN
711 Third Avenue, New York, NY 10017, USA

*Routledge is an imprint of the Taylor & Francis Group,
an informa business*

First issued in paperback 2016

Typeset in Sabon by Keyword Group Ltd

British Library Cataloguing in Publication Data
A catalogue record for this book is available from the British Library

Library of Congress Cataloging in Publication Data
Styhre, Alexander.
Managing knowledge in the construction industry / Alexander Styhre.
  p. cm.
  Includes bibliographical references and index.
    1. Construction industry–Information resources management.
    2. Knowledge management. 3. Construction
    industry–Communication systems. 4. Construction
    workers–Training of. 5. Architects and patrons. I. Title.
  TH215.S78 2009
  690.068'4–dc22   2008038091

ISBN 13: 978-0-415-46344-7 (hbk)
ISBN 13: 978-1-138-99556-7 (pbk)

Spon Research ISSN: 1940-7653
Spon Research EISSN: 1940-8005

Contents

Tables

Preface

As all practising researchers know, it is in many cases very complicated to determine *ex post facto* when a research project starts and ends. There are naturally starting and termination dates for projects being granted research funding, but in day-to-day work things tend to get more complicated. At times, the reading of a paper, meeting with a person or some other memorable event may be regarded as the starting point for a particular endeavour, but in most cases the memory is just of a vast number of days filled with reading, writing, interviewing and other similarly tedious and/or insignificant activities filling up the social science researcher's everyday work life. This book is the outcome from one such long-term research project, devoid of an exact starting point and – as it may turn out – a proper ending.

As academic researchers in the management disciplines, we are at times critical of colleagues in industry who fail to evaluate their activities properly and systematically. But we often do not practise what we preach, and are left with only a series of scattered memories and a few notes to collect and account for when the empirical material is being used later on. What I *do* recall, however, is that I have been interested in the concept of knowledge for a long time, starting in the mid-1990s when I did my Ph.D. thesis on the application of the then-fashionable Japanese management principles in Swedish manufacturing industry. When I began conducting research in the construction industry at the beginning of the new millennium, the concept of knowledge seemed to lurk in the background every now and then. As a consequence, in 2006, I was granted research money for a project on the management of knowledge in the construction industry. This book is the outcome of that project, but it is also strongly influenced by previous research work. In my mind, this book is an attempt at summarizing a series of projects and bringing them together under the umbrella of knowledge management.

As always, the author of the book is heavily indebted to a number of people. Hans Trulsson, PEAB, and Ronald Caous have strongly influenced the study of the coaching of site managers. Lars-Göran Dahlquist, the CEO of ConCo, helped me arrange interviews in the company and provided valuable insight into the world of rock construction workers. Fredrik Nilsson at Chalmers University of Technology and Brown Architects provided invaluable help in the study of the work at the architect's office accounted for in Chapter 4. Pernilla Gluch and Per-Erik Josephson at Chalmers University of Technology have served as my colleagues and discussants in the knowledge management research project. I would also like to thank the members of the scientific board of the Center for Management in the Construction Industry (CMB) at Chalmers University of Technology for insightful comments on the order of things in the construction industry. My friend and former colleague Mats Sundgren conducted most of the data collection in the architecture work study, and also contributed with helpful comments and remarks on the text in various drafts. Some of the ideas presented in this book were presented at the seminar at Scancor, Stanford University, in October 2007, and I am grateful for comments from Ester Barinaga, Claes Bohman, Geerte Hesen, Maria Jarl, Erik Piñeiro, Anne Reff Pedersen, Eero Vaara, Nina Veflen Olsen, and Karl Wennberg. I am also grateful for valuable comments on the concept of aesthetics from Erik Piñero. In addition, I would like to thank all those construction industry representatives who have in various ways contributed with their insights, experiences, life stories or opinions to this book. Finally, a thank you goes to my family, Sara, Simon and Max, for being around in good times and bad.

Alexander Styhre
Gothenburg, August 2008

Introduction

Entering the knowledge society

To ward off potential future criticism and disappointed readers, it deserves to be pointed out right away that this monograph is not intended to serve as a 'how to' book in the field of knowledge management; it is not anchored in the engineering sciences with their insistence on solving the problems at hand. Instead it is an attempt on the basis of a variety of social sciences critically to discuss and examine how knowledge can be addressed, managed and developed in the construction industry. The literary corpus addressing the management of the construction industry tends at times to enact an instrumental and functionalist perspective, thereby reducing inherent complexities to linear relationships and uncomplicated facts of the matter. Such a perspective is enormously rewarding in terms of bracketing off the full complexity of social and ordinary life, and focusing exclusively on solving pressing problems. However, operating exclusively on the basis of what has been called 'downstream' theory (Nayak, 2008) eliminates some of the more elementary assumptions within a discipline or field of investigation. Therefore, this book has the ambition to think not only of 'knowledge management' as a fixed or uncomplicated set of practices, models, concepts and tools, but to think equally of 'management' as a social practice and 'knowledge' as an epistemological category as embedded in social, economic and cultural relations that strongly shape and form how these terms are used. That is, 'knowledge management' does not fall from the sky but instead denotes a series of social practices in organizations that in various ways are contingent on historical and social contexts. Taking such an 'anti-essentialist' view of knowledge management is a complicated task because it does not assume that there is some transcendental idea or commonly shared model of what knowledge management is prior to actual practices. Instead, knowledge management becomes the emergent practice wherein various forms of skills and know-how are treated as an organizational resource that is contributing to the firm's long-term competitiveness and sustainable competitive advantage. Given the substantial heterogeneity of the forms of knowledge mobilized in the construction industry – ranging from the architect's vision of how social spaces can be transformed into build environments to the carpenter's ability to use various mechanical tools to produce actual buildings – knowledge management in the construction industry will of necessity become a rather amorphous term. It contains a wide range of activities, practices, tools, procedures and systems.

This book addresses the management of knowledge in the construction industry. This declaration immediately calls for a more thorough examination of highly malleable – some would say fuzzy – terms such as 'manage', 'knowledge' and 'construction industry'. The concept of management, whose etymology enables diverse interpretations and meanings, may include terms such as 'control', 'guidance', 'surveillance', 'direction' and so forth; knowledge is a term that includes a whole set of cognitive, embodied and emotional skills, capacities and insights; the construction industry is constituted by a multiplicity of professions, occupational groups, firms, corporations and enterprises, mobilizing and using various aesthetic, symbolic and material resources. In other words, to claim that one will discuss how to manage knowledge in the construction industry is either to assume that the reader shares a significant number of assumptions and beliefs, or is wholly ignorant of the ambiguities involved when making such a declaration. Anyway, one must not be overwhelmed by the complexity of linguistic resources; most practical things do work fine when all the theoretical and epistemological intricacies are bracketed and ignored. The point is that the management of knowledge in the construction industry is too vast a subject to be addressed in a single volume of a research monograph. Rather than taking on the burden of capturing the very essence of knowledge management in this particular industry (in a sense a 'grand theory' project as persuasively refuted by Charles Mills Wright in his *The Sociological Imagination*), the book seeks to examine some relevant aspects pertaining to the use of intellectual resources in building and architecture work. At the same time, as scholars of science and technology and actor–network theorists teach, neither nature nor society speaks for itself. Any 'empirical entity' (e.g. an observation, an interview excerpt) is theoretically 'overdetermined'; it can be examined in many different ways (Becker, 1992). Therefore, the empirical studies reported in Chapters 2–5 are accompanied by theoretical frameworks that enable a more detailed analysis of the empirical material. In so doing, the book attempts to navigate in the space between construction engineering books providing advice and recommendations on how to manage the construction project, and the more interpretative and analytical literature in the field of organization theory in general and knowledge management more specifically. The book does not seek to formulate answers to questions, but rather grapples with the questions themselves, and thereby hopefully is capable of pointing to some useful aspects of the management of knowledge in the construction industry.

The term 'knowledge management' is a product of the 1990s and the substantial growth of jobs in the technical and professional sectors at the expense of, primarily, the manufacturing industry. For instance, Robertson and Swan (2004: 129) point out that there is little doubt that sectors characterized by 'knowledge work' are growing, and they report that the science and technology sector has grown between 4 and 16 per cent annually over the last 15 years. Barley and Kunda (2006: 55–6), speaking about the United States (US) economy, defuse the myth that clerical and service work has mitigated the decline in manufacturing jobs. Clerical work peaked in 1970 at 18 per cent of the workforce and has subsequently declined by a percentage point. Service employment has only grown by about 4 per cent since 1960 and accounts for about 16 per cent of the US workforce. In addition, managerial positions and sales work have grown by 1.5 and 4 per cent, respectively, since 1950. Instead it is white-collar employment that stands for the largest growth in the US economy: 'Since 1950 professional and technical employment more than doubled, growing from 8 percent to 18 percent of the workforce. In fact, by 1991, professional and technical workers had become the largest sector surpassing even clerical workers and operatives', Barley and Kunda (2006: 55–6) conclude. Frank and Meyer (2007: 289) point at the explosive growth of university education in virtually all parts of the world, another indication of the alleged 'knowledge society': 'In 1900, there were about three tertiary education students per 10,000 worldwide. By 1950, this number had increased eight-fold to 25. By 2000, it had increased another six-fold to 166'. In a hundred years, the number of university students per capita grew 55 times.

Powell and Snellman (2004) examine patterns in patenting in the US and Europe to find evidence for what they call the 'knowledge economy'. They report that, even though the period from 1963 to 1983 did not show any particular growth in patents, after 1987 there is a substantial growth in registered patents, from 80 000 to 170 000 patents annually. 'Clearly', Powell and Snellman (2004: 202) contend, 'patent trends suggest a recent marked acceleration in the production of new knowledge'. Powell and Snellman (2004: 204) also make reference to a study published by Jason Owen-Smith finding an eightfold increase in university patents in the period 1976–98. Moreover, Powell and Snellman (2004: 205) found a strong increase in non-academic science and engineering (S&E) jobs between 1980 and 2000. S&E employment increased by 159 per cent in the 20-year period, corresponding to an annual growth of 4.9 per cent, in comparison to 1.1 per cent growth in the entire US labour force. In summary, there is ample evidence that today's labour force is more specialized and conducts more advanced and knowledge-intensive work than in the 1970s and earlier. Powell and Snellman (2004) reserve the term 'knowledge economy' to denote this new economic regime that is under way:

> We define the knowledge economy as production and services based on knowledge-intensive activities that contribute to an accelerated pace of technological and scientific advance as well as equally rapid obsolescence. The key components of a knowledge economy include a greater reliance on intellectual capabilities than on physical inputs or natural resources, combined with efforts to integrate improvements in every stage of the production process, from the R&D lab to the factory floor to the interface with the customer.
>
> (Powell and Snellman, 2004: 201)

However, the knowledge economy, and as Knorr Cetina (1997: 8) argues, the *knowledge society* is not only a society of simply more experts, of technological infra- and informational structures, and 'of specialist rather than participant interpretations', but is a society where 'knowledge cultures' have 'spilled and woven their tissues into society, the whole set of processes, experiences and relationships that wait on knowledge and unfold in its articulation'. By this formulation, Knorr Cetina (1997) suggests that knowledge is not 'additional to' or optional in contemporary society, but that the knowledge society evolves on the basis of the very production and circulation of knowledge; knowledge is infrastructural rather than ornamental – it is the very fabric of society rather than its *Über-bau*. In such a society and economy, essentially based on intellectual capital (here used as a formal concept denoting a stock of know-how), the concept of knowledge management quickly becomes a distinct research domain in the field of management research. The term today denotes a rather diverse and heterogeneous field of research, sharing the basic assumptions that it is knowledge that is the single most significant factor when explaining differences in performance between different firms and industries. While knowledge management has primarily focused on traditional knowledge-intensive sectors of the economy such as new product development and innovation work, and various domains of professional work, the construction industry has been (perhaps) surprisingly little examined from this theoretical perspective. One of the explanations may be that while, for instance, manufacturing industry and innovation work have been dominated by engineers, a professional group established since at least the few last decades of the nineteenth century (Shenhav, 1995, 1999), construction firms tend to hire occupational and semi-professional groups of workers and engineers whose status as knowledge workers are somewhat ambiguous. Already Stinchcombe (1959) had emphasized that construction work has maintained a craft-like production form (a position thoroughly criticised by Eccles, 1981) as opposed to manufacturing industry, which was essentially restructured during the 'rationalization movement' orchestrated by the emerging group of professional engineers (Guillén, 1994; Shenhav, 1995). Therefore, speaking of construction companies as being 'knowledge-intensive' is problematic if that term is reserved to denote the work of, say, lawyers, laboratory researchers or mechanical engineers. At the same time, such a declaration calls into question what the very term 'knowledge' is supposed to or may denote: is it a term reserved for prestigious professions (e.g. medical doctors and university professors) or can it be used to capture any social practice that mobilizes a certain degree of specialized knowledge in its pursuits? In this book, knowledge is used pragmatically as a portmanteau term denoting all sorts of cognitive, emotional and embodied skills and capacities that are used in a social practice. Knowledge is then not a term privileged with mythical or extraordinary qualities but is instead a 'knowledge of everyday life'.

The size and impact of the construction industry

To start off, a formal definition of the construction industry is provided:

> The construction industry comprises all those organizations and persons concerned with the process by which building and civil engineering works ... are procured, produced, altered, repaired, maintained, and demolished. This includes companies, firms, and individuals working as consultants, main and sub-contractors, material producers, equipment suppliers and builder's merchants. The industry has a close relationship with clients and financiers.
>
> (Hillebrandt, 2000: 4)

Eccles (1981: 451) emphasizes four characteristics of the construction industry: (1) it has a 'small degree of diversification', that is, construction firms deliver approximately the same products and services; (2) construction firms operate on 'geographically limited markets', often in just one country or region; (3) there are relatively low entry barriers; and (4) there is a 'lack of concentration' in the industry, that is, there are large numbers of actors in the industry.

A knowledge-management perspective is of relevance for the construction industry, if not in terms of its inherent qualities and skills, then in terms of its sheer weight and influence in the economy. The construction industry accounts for about 10 per cent of gross domestic product (GDP) in the world as a whole (Hillebrandt, 2000: 19). In the UK, the construction industry is the largest sector in terms of both its proportion of GDP and the number of people it employs (Agapiou, 2002: 697). In the UK, the construction industry employs 1 169 000 people, of which 154 500, or 13.2 per cent of the workforce, work in the largest firms employing more than 1200 people (Construction Statistics Annual Report, 2006: table 3.4, p. 50). In total there are 182 644 registered construction companies in the UK, but only 56 firms hire more than 1200 employees (up from 33 in 1995). The entire industry has an annual turnover of £22 654 million, of which £4489.4 million, or 19.8 per cent, is accounted for by the largest firms (Construction Statistics Annual Report, 2006: tables 3.1 and 3.4, pp. 45, 48). In Sweden, the construction industry is the second largest sector of the economy, outnumbered only by the large public health care sector. The construction industry is a major industry in all Organisation for Economic Co-operation and Development (OECD) countries, and employs a great variety of professional and occupational groups. In addition, the construction industry is labour intensive in comparison to, for instance, manufacturing (Hillebrandt, 2000: 187), and what economists call *multiplier effects* generate additional work in other industries and sectors of the economy. It is therefore tempting for governments to use the construction industry to 'manage the level of demand in order to reduce short-term fluctuations in the economy' (Hillebrandt, 2000: 187). In general, the construction industry is susceptible to swings in the business cycle and therefore economic fluctuations are endemic in the industry (Hillebrandt, 2000: 26).

In addition to its sheer size and importance for global, national and regional economies, there are numerous studies pointing at the lower productivity growth and a limited degree of innovation in the construction industry (Dorée and Holmen, 2004; Harty, 2005).[1] Gann provides some figures emphasizing the comparatively lower productivity growth in construction:

> Lower rates of productivity growth in construction compared with manufacturing have contributed to a relative increase in construction costs ... data suggests that construction has failed to keep pace with performance improvement realized in other sectors. In the period 1970 to 1985, productivity in European construction increased at an average of 0.9 per cent per annum which was low in comparison with other industries ... Construction in a number of countries including the United States, experienced negative productivity growth [in 1970–93]. This compares with labour productivity growth of between 3 per cent and 4 per cent annually between 1985 and 1995 in high and medium-high technology industries.
>
> (Gann, 2000: 6)

Concerning innovative capabilities, Drejer and Vinding (2006: 928, n. 1) reports a survey study in Denmark showing that, while 58 per cent of the firms in the manufacturing industry and 44 per cent of the trade and services firms in the study had introduced new products or services during a particular period, the corresponding figure for the construction industry was a meagre 22 per cent. The relatively weak innovative capabilities of the construction industry have been explained by its relatively fragmented industry structure, organized into loosely coupled networks of firms (Dubois and Gadde, 2002), the project organization (Drejer and Vinding, 2006), or more generally in terms of a general scepticism towards new materials and techniques. Moreover, Agapiou (2002) points at a number of problems facing the construction industry: 'It is widely accepted that the UK construction industry often falls short of meeting the needs of its clients, the developers, in terms of construction times, costs, predictability, quality, defects, safety, waste minimization' (Agapiou, 2002: 698). In the widely discussed Egan report (1998), commissioned by the British Construction Industry Council, it is stated that 'The industry as a whole is underachieving ... Too many of its clients are dissatisfied ... projects are widely seen as unpredictable in terms of quality on time, within budgets and to the quality expected' (cited in Agapiou, 2002: 698). In another official report, it was found that '73 per cent of projects over tender prices; 70 per cent deliver late' (Report on Modernizing Construction, National Audit Office, UK, cited in Agapiou, 2002: 698).

Taken together, the construction industry is in great need of optimizing the use of its intellectual resources. In addition, since the built environment is what strongly influences everyday work-life in contemporary society, and the cost of living and housing accounts for a substantial part of the private and public economy, the ability to exploit existing bodies of know-how more effectively is a widely desirable objective.

Purpose and outline of the book

Given the strong emphasis on the production and circulation of knowledge in today's society, and the importance of the construction industry, the purpose of the book is to discuss knowledge-management practices within this industry. Rather than assuming it is possible to formulate a unified, universally applicable theory about 'how to manage knowledge in the construction industry', the book looks at three different research projects in the construction industry over the period 2004–9. The studies include research in both conventional construction firms and in architecture bureaus. Moreover, the four empirical chapters of the book (Chapters 2–5) are not structured in accordance with a single integrated analytical model; instead, a heterogeneous body of literature is used in each chapter, applicable in individual studies. Expressed differently, it may be argued that the approach to the four individual studies is synthetic rather than analytical; the chapters do not seek in the first place to provide an answer to the question 'how to manage knowledge', but to show how local and contingent practices in different domains of the heterogeneous construction industry demand different approaches and need to be examined from alternative theoretical perspectives. However, this does not mean that the book is wholly devoid of managerial implications and suggestions for practices pertaining to knowledge work in the construction industry. In the final chapters, some of the arguments and findings are summarized and further discussed. The book is thus structured accordingly.

In Chapter 1, the analytical framework for studying the use of knowledge in the construction industry and companies is developed. First, occupational and professional groups define and structure what are legitimate and useful forms of knowledge in a particular field of expertise. In the construction industry, a wide number of occupational groups are involved in the work, ranging from architects and designers to carpenters and electricians. An understanding of how occupational groups conceive of their roles, duties and privileges is very helpful for gaining insight into how knowledge is mobilized and used *in actu* and *in situ*. Next, the sociological concept of practice is introduced, serving as an intermediate analytical level between occupational and professional groups and everyday practice. The concepts of knowledge and knowledge management, central to the whole book, are then discussed. Thereafter, the literature on knowledge management practices in construction industry is reviewed.

In Chapter 2, a study of how executive coaching can be used to support and help site managers in major construction projects is reported. The chapter is based on a three-year study of the changing conditions for site managers' work and how executive coaching can be used as an approach to help site managers handle all their day-to-day assignments.

In Chapter 3, a study of work in architecture, a quintessential knowledge-intensive domain of the construction industry is presented. Making a comparison with scientific laboratory work, the everyday work-life of practising architects is conceived of as sharing a basic morphology of work with laboratory scientists. In this chapter, a body of literature commonly known as 'science and technology studies' is referenced.

In Chapter 4, which reports another study of an architect's office, the concept of visual artefacts and what has been called 'professional vision' are examined. Conceptualizing professional vision in terms of what Jacques Lacan calls 'the gaze', architects' professional know-how is embodied and part of their ability visually to inspect and conceive of possible solutions to perceived problems and challenges.

In Chapter 5, a study of a specialized construction company, ConCo, using expert know-how in rock construction, including techniques such as spray concrete and rock injection, is reported. In this chapter, the concept of social capital is invoked to understand how site managers in the construction company are capable of sharing knowledge on a day-to-day basis through verbal communication and relatively mundane media, such as telephones and a few diary notes from individual projects.

In Chapter 6, the final chapter, a few concluding remarks are made and some practical implications addressed. The book finishes with a few suggestions for further research and alternative routes to explore in how to manage knowledge in the construction industry.

1 Occupational groups and professions, practices, institutions and knowledge in construction work

Introduction

The construction industry is constituted by a plethora of social practices and materialities, and involves a long series of occupational and professional groups (Bowen, Pearl and Akintoye, 2007). In order to understand how knowledge is formed, articulated and circulated in such heterogeneous environments, a number of analytical concepts need to be brought into discussion. In this chapter, some of the central concepts of the theoretical framework guiding the empirical studies are introduced and discussed. Such concepts include *occupational and professional groups*, *practice*, *routines*, *rules* and *standard operation procedures*, and they form a theoretical framework underlining the actual day-to-day practices in construction projects as the constitutive components of any knowledge-management initiative in the industry. To put it differently, the perspective taken in this book is essentially 'bottom-up', emphasizing the everyday work procedures in knowledge work rather than a 'top-down', information-management perspective where knowledge management systems are defined a priori and practices are located within a determining system. In the latter half of the chapter, the literature – or rather *some* of the literature belonging to the fast-growing corpus of texts – addressing the intersection between knowledge and management is reviewed. The chapter thereafter addresses some of the central characteristics of the construction industry, pointing towards the more empirically oriented parts of the book.

Professions, occupations and practice: analytical tools for understanding construction industry work

In order to examine how various forms of knowledge are mobilized in the construction industry, a few analytical tools are needed. In the following, two theoretical domains will be examined in some detail.

First, the concepts of occupations and professions are elaborated upon and thereafter the concept of practice, here conceived of as an analytical category, is discussed.

Occupations and professions: sorting out and demarcating terrains

Studies of occupational identities and professions represent a classical field of research in sociology and neighbouring disciplines in the social sciences (Illich, 1977; Larson, 1977; Freidson, 1986). Professional groups have organized themselves into guilds and other professional communities since at least the middle-ages (Braudel, 1992), and with the emergence of a modern society, professional identities were even more accentuated. In the modern period, professions have been defined and credentialized by the state or organizations – Freidson (1986) here talks about 'institutional credential-izing' – being given the legitimate right from the state to organize the production of particular kinds of services to the public, including 'training or education or prospective members of an occupation' (Friedson, 1986: 64). In addition, professional expertise and legitimacy have been strongly tied to the institution of the modern university, the predominant institution in contemporary society producing, regulating and controlling systematic and scientific knowledge.

However, for emerging professions that has not always been the case. Prior to the modernization of universities and the establishment of proper scientific procedures, Larson (1977) argues that, in some cases, the university actually hindered rather than helped the production of systematic scientific and technical knowledge. For instance, in the case of medicine, hindered in its development by incumbent doctrines and beliefs at the universities and various guilds' claims on jurisdiction, in the nineteenth century, new medical practices were largely developed at the hospitals, amidst the everyday medical practice. Larson (1977: 24) argues that one of the main reason for Paris being the world's capital of medical science during the first half of the nineteenth century was its large number of hospitals and that these hospitals were bringing surgeons and physicians together, thereby overcoming the *ancien régime* of guild barriers. When surgeons and physicians collaborated, physicians started to incorporate the localized structural pathology that surgeons had spontaneously applied in the scientific study of specific diseases. The emergence of modern professions and professional authority and credentializing systems is one example of the effects brought by organizational capacities. Larson (1977) emphasized this organized nature of professions:

> [T]he professional project is an organizational project; it organizes the production of producers and the transaction of services for a market; it tends to privilege organizational units in the system of stratification;

it works through, and culminates in, distinctive organizations – the professional school and the professional association.

(Larson, 1977: 74)

Professions are thus defined, somewhat simplified, by their ability to monopolize specific domains of expertise, a definition that Attewell (1990) would refer to as being Weberian in term of emphasizing the struggle over power and prestige in a particular field rather than the nature of expertise *per se*. Herein lies also an important difference between professions (e.g. lawyers and medical doctors, the two most clear-cut and conventional cases, but also more 'fuzzy' professions, such as engineers or business school graduates) and the occupations. For instance, occupational groups do not of necessity make use of less esoteric or specialized knowledge and expertise (think, for instance, of a watchmaker or a midwife) than professional groups, but they belong to an occupational group less successful in defending and monopolizing the jurisdiction over their domains of expertise. The professional status is in this Weberian view ultimately a matter of power and the ability to establish monopolies or at least significant entry barriers. In the 'grey areas' where professional and occupational groups collaborate, there is a strong emphasis on what Gieryn (1983) calls the 'boundary work' between the two categories – the longstanding struggle over authority and the right to conduct certain operations among obstetricians and midwives is a well-known and representative case – safeguarding the authority of the privileged professional group. However, the distinction between professional and occupational groups is not a binary one, but is to be examined along a continuum ranging from the highly monopolized profession with high entry barriers (e.g. medical doctors) to occupational groups with relatively low or non-existing entry barriers (e.g. taxi-drivers or waitresses). [For a formal categorization of occupational and professional groups, see United States (US) Office of Personnel Management's (1998) list of occupational groups.] Thus, studies of how professional groups define, develop and monitor their domain of expertise are also of relevance for occupational groups. Expressed differently, when taking away some of the specific features of the work of professional groups (see Attewell, 1990: 437–8), professional and occupational groups are defined on the basis of their social status.

In Hughes's (1958) seminal text *Men and Their Work*, six different ideal-types of rationality guiding occupations are identified: (1) those guided by a mission, for instance, to engage in religious teaching or work to help the poor and needy; (2) professions and 'near-professions', that is, occupations sanctioned by the state authorities, for instance, medical doctors or lawyers; (3) enterprise, dealing with commodities; (4) arts; (5) trades – 'very close to the arts', Hughes (1958) remarks – and finally, (6) jobs. In this taxonomy, individuals engaging in various activities go from the fully committed and dedicated (as in the case of the religious preachers endowed with a calling) to a more detached attitude, primarily regarding the job as a source of income but little more. As Hughes (1958: 75) remarks, no line of work cannot be fully understood outside of the 'the social matrix in which it occurs or the social system of which it is part'. That is, the system includes not only the recognized institutional complex of the occupation but also 'reaches out to and down into other parts of the society'. Therefore, occupations and professions always contain ambiguities and 'apparent contradictions in the combinations of duties' (Hughes, 1958: 75). That is, occupations and professions are negotiated social orders depending on many intersecting factors and conditions. For instance, in the study by Strauss, Schatzman, Bucher *et al.* (1964) of professions and occupational groups in the psychiatrist health care sector in the US, professional groups are conceived of as the outcome from an 'emergent process' bound up with ideologies and treatment practices; 'Specializations are anything but stable entities with relatively fixed boundaries and tasks', Strauss, Schatzman, Bucher *et al.* (1964: 6) state. Professions and professional attitudes and beliefs are therefore never given as such but are instead the outcome from pre-existing ideologies that individuals converge to. The concept of ideology is here denoting an 'abstract system of ideas' that is mediated by 'operational philosophies'. The operational philosophies are in turn 'systems of ideas and procedures for 'implementing therapeutic ideologies under specific institutional conditions' (Strauss, Schatzman, Bucher *et al.*, 1964: 360). That is, ideologies are the overarching system of beliefs guiding day-to-day work, and the operational philosophies are the more down-to-earth actual practices engaging material resources. Studies of professional and occupational groups such as policemen and policewomen (Van Maanen, 1975), managers (Dalton, 1959; Jackall, 1988), fast-food restaurant workers and salesmen (Leidner, 1993), restaurant chefs (Fine, 1996), restaurant waitresses (Paules, 1991), copy-machine repair technicians (Orr, 1996) and manufacturing workers (Roy, 1952; Burawoy, 1979) suggest that ideologies, beliefs and norms guide and structure everyday work and set the boundaries for what qualifies as being legitimate work. However, studies of, for instance, police work (Jermier, Slocum, Fry *et al.*, 1992) show that there is substantial leeway between what policemen do and formally enacted procedures for police work. Under slogans such as 'serve and protect' or 'crime-fighting', there is substantial space for individual and local translations of such objectives into actual performances. Such interpretations are socially embedded, i.e. 'ideological', according to Strauss, Schatzman, Bucher *et al.* (1964). Lamont and Molnár (2002: 178) discuss the study of Collins (1979) who found a 'surprisingly weak correlation between the requirements of educational credentials and the skills/knowledge requirements of jobs':

> Education is often irrelevant to on-the-job productivity, and is sometimes counterproductive. Specifically vocational training seems to be derived primarily from work experience rather than from formal

school training. The actual performance of school themselves, the nature of the grading system and its lack of relationship to occupational success, and the dominant ethos among students suggests that schooling is very inefficient as a means of training for work skills.

(Collins, 1979: 21)

On the basis of this empirical observation, Collins (1979) argues that education serves to 'socialize prospective professionals into status cultures by drawing a line between insiders and outsiders'. That is, one does not primarily attend tertiary schooling to learn practical skills but to be trained at thinking and behaving as a member of a particular social group. In fact, Larson (1977: 226) goes so far as to argue that professions are more often defined as being an occupation which tends to be 'colleague-oriented' rather than 'client oriented'. For instance, university professors tend to be more concerned about how colleagues, and especially leading researchers, regard their scientific contributions than how students perceive and evaluate their teaching. Similarly, Murningham and Conlon (1991) found in their study of 20 professional British string quartets that string quartets were more inward-oriented than oriented towards the audiences when seeking to accomplish musical performance at the peak of their capacity. In order fully to evaluate and appreciate the skills of the professional, you need to be a member of the professional community; professionals always and of necessity appreciate esoteric knowledge. It is part of their training, socialization and enacted ideology.

Several studies also show that professional and occupational ideologies are what are accommodated during secondary schooling (Willis, 1977), or in professional education and training in university programmes (Becker, Geer, Hughes and Srauss, 1961; Johnson, 2007). Schleef (2006), studying law and business school students – future 'managing elites' in Schleef's (2006) view – argues in line with Collins (1979) that becoming a professional is a processes wherein the student must actively resist elite ideologies in order to accommodate them. Rather than being passive recipients of predominant professional ideologies, students are trained to think critically and to question assumptions. In Schleef's (2006) view, students are not 'unwilling dupes of ideological indoctrination' but are self-reflective and capable of strategically accommodating and resisting ideologies of their education. Managing elites need to accommodate these ideologies because, in their future professional work, they 'need to believe in the higher mandate that the professionals are alleged to embody', Schleef (2006: 5) says. 'Elites-in-training' therefore undergo, Schleef (2006: 4) argues, a process where they 'contest, rationalize, and ultimately enthusiastically embrace their dominant position in society'. For instance, Danielle, a law school student who 'firmly believed during her first year of law school that most lawyers were overpaid and took advantage of their powerful position in society', now says, without criticism: 'Lawyers work really,

really hard ... the money is deserved. I think lawyers are really, really smart. I think they are very articulate and on top of things' (Schleef, 2006: 2). One of the principal mechanisms for this accommodation of elite ideologies is what Schleef calls *surface cynicism,* manifested in, for instance, a scepticism towards the value of formal training – 'it is all common sense', the first year business schools students, claimed – the pedagogical tools in use in the training (e.g. the use of the so-called 'Socratic method' in law schools), and the position of elites in society more generally. In Schleef's (2006) account, surface cynicism serves a functional role in providing a form of symbolic resistance that helps the students becoming professionals:

> Surface cynicism is a symbolic resistance that creates and strengthens elite solidarity. Students unite against the elements of their schooling that they can reject, in order to show that they have not been too taken in by school rhetoric and do indeed see behind the façade of professional ideology. At the same time, the dynamics of student resistance actually fortify many aspects of professional ideology and cause students to become more intricately invested in their disciplines ... Criticism of school is an expected part of the student persona, but total rejection or acceptance of law school rhetoric is not. Students can recognize and critique messages about the pedagogy without jeopardizing their investment in the professional hierarchy.
>
> (Schleef, 2006: 91)

However, after a period of training, the students start to recognize the value of 'thinking like a lawyer' or being capable of 'strategizing like a manager', and gradually take on a professional persona. Still, they maintain that they are capable of 'seeing through the guise of school message' (Schleef, 2006: 203), thereby telling themselves that some autonomy from the dominant ideology of schools, recruiters and society is possible. Schleef's (2006) study shows that becoming a member of the managing elite is a delicate balancing act between a rejection and a passive submission of the dominant ideologies and institutions of society. Professionals are trained to think autonomously and put things into question, yet to be capable of adhering to predominant beliefs, and therefore students actively have to engage in a continuous reconstruction of who they are and what they want to be. It is little wonder then that most of the students interviewed by Schleef insisted on being 'different from the other students' and that they would 'skew the sample' in the study, at the same time as they were eager to hear whether their views conformed or diverged from that of the other students.

In the account by Strauss, Schatzman, Bucher *et al.* (1964), it is between this abstract system of beliefs and the operational procedures that professional identities emerge. They identified three dominant ideologies among psychiatry professionals: (1) a *somatic ideology,* where psychiatrists were committed to 'organically based etiology and procedures', that is, the use of medication; (2) the *psychotherapeutic ideology* where various forms of psychotherapy were endorsed; and (3) the *milieu therapy ideology* underscoring the 'crucial importance of environmental factors in etiology or treatment' (Strauss, Schatzman, Bucher *et al.*, 1964: 8). These different ideologies guided and structured not only the professional groups but also structured the operational procedures in the various hospitals examined: 'Ideologies provide frameworks for judging both how patients should be helped and what is harmful for patients' (Strauss, Schatzman, Bucher *et al.*, 1964: 365). Thus, acquiring beliefs regarding what constitutes 'proper psychiatrist work' is an integral component of professional training:

> While the future professional is engaged in acquiring the specific skills of his trade and the professional identity that will guide his activity, he also acquires convictions about what is important or basic to treatment and what is proper treatment, he learns treatment ideology as an integral part of his professional training.
>
> (Strauss, Schatzman, Bucher *et al.*, 1964: 363)

Not only are professional ideologies affecting how actual practices are carried out, they are also imposing certain moral sentiments on the individual. The 'professional canons' are, as Strauss, Schatzman, Bucher *et al.* (1964: 365) put it, 'highly charged morally'. This means that professional ideologies were strongly affecting how individuals perceived a situation and how they behaved under specific conditions. Strauss, Schatzman, Bucher *et al.* (1964) emphasized what has been called 'normative isomorphism' (DiMaggio and Powell, 1983; Mizruchi and Fein, 1999) in psychiatry work:

> The fieldwork data suggests that institutions are both selective and productive in terms of ideologies. They are selective in that only certain types of ideology can be tolerated or implemented within the limits set by both institutional necessities and the particular organization of treatment. For example, on the state hospital treatment services, any young psychiatrist whose ideological orientation was basically psychotherapeutic had to develop a scheme of operation drastically modifying the psychotherapeutic approach appropriate in other institutions.
>
> (Strauss, Schatzman, Bucher *et al.*, 1964: 360)

At the same time, psychiatric institutions are by no means closed or 'total institutions' (Goffman, 1961) but are continuously under the influence of outside communities and objectives. Given that, professional ideologies are always open to modifications and reformulations while they are capable of preserving a core of beliefs over time. In general, professional ideologies are systems of fairly persistent beliefs and assumptions that guide and structure an individual's day-to-day work. The study of Strauss, Schatzman, Bucher *et al.* (1964) shows that professional ideologies may tolerate some deviances from the norm, but that they also tend to make individuals converge towards legitimate and widely endorsed beliefs and practices. In other words, professional identities are emerging properties dependent on a variety of social conditions and events. In addition, the professionalization and socialization into occupational groups are never fully accomplished but are continuous and ongoing processes. Being a member in a particular organizational community means that one is adapting to specific ways of talking and behaving:

> Individuals in an organization field undergo anticipatory socialization to common expectations about their personal behavior, appropriate style of dress, organizational vocabularies ... and standard methods of speaking, joking, and addressing others.
>
> (DiMaggio and Powell, 1983: 153)

As suggested by Strauss, Schatzman, Bucher *et al.* (1964), the status of professionals is always reproduced through the adherence to enacted norms and behaviours including embodied and verbal activities. On the other hand, professionals not only respond to pre-existing institutions and converge towards standardized scripted behaviour but are, as Scott (2008) underlines, themselves primary 'institutional agents', agents contributing to the establishment of new institutions. 'More so than any other social category', Scott (2008: 223) writes, 'the professions function as institutional agents – as definers, interpreters, and appliers of institutional elements. Professionals are not the only, but are – I believe – the most influential, contemporary crafters of institutions'. Seen in this view, there is a complex recursive relationship between individual professional taking on professional behaviours and symbolism (e.g. dress codes, the use of enacted vocabularies and phrases) and professional communities as active institutional agents establishing new institutional behaviours and conditions.

More recently, professional ideologies emerge from new economic and social conditions. Barley and Kunda's (2006: 55) study of computer programming professionals in the San Francisco Bay Area's Silicon Valley computer industry introduces the term 'itinerant professionalism'. These itinerant professionals are highly educated and skilled 'hired guns', programmers that have chosen to operate on a freelance basis in the industry. For Barley and Kunda (2004), the concept of a professional group is still bound up with clear lines of demarcation between insiders and outsiders:

> Professionalization is typically associated with an occupation's possession of an esoteric body of knowledge, state-mandated licensing,

formal training programs, and professional associations that create barriers to entry and a basis for defending their jurisdiction from the expansionist tendencies of other occupations.

(Barley and Kunda, 2004: 294)

One important part of any professional group's work is to engage in what Gieryn (1983), speaking of the line of demarcation between science and 'non-science' (e.g. pseudo-sciences such as phrenology), names boundary work (see also Bechky, 2003a). Boundary work is serving to 'monopolize professional authority' in a specific domain of expertise and thus serves an important role to exclude certain groups or individuals for the benefit of individuals qualifying as members of the professional group. For instance, medical doctors are given authority in the field of medicine and health care through acquiring a university diploma from a medical school and through practising in the health care sector. Medical doctors are eager to blacklist any attempts to lower their professional status and therefore alternative medicine cannot be tolerated as a legitimate source of health care. At times, as for instance in medicine, the boundary work is clear and rather unambiguous, while in many domains of expertise it is more complicated for outsiders to understand the mechanisms of boundary work. For instance, in Livingston's ethnography of mathematicians it is not the ability to undertake a specific calculation that constitutes one's status as a mathematician but the ability to formulate 'proofs' of mathematical theorems:

> For mathematicians, the heart of their profession is not theorem proving *per se,* that it is itself a classical version of their work ... the sustaining life of professional mathematicians lies in what might be called 'mathematical structure building' – that is, the envisionment and construction of structures of theorems and proofs that have their motivating origin in, and are directed to the development and reformulation of, a current state of mathematical practice ... without knowing, as praxis, the work of theorem proving, the discovery, construction, recognition and consequentiality for mathematical practice of such mathematical structure is impossible.
>
> (Livingston, 1986: 177–8)

A mathematician does not need only to be capable of using mathematics but is expected to understand fully the underlying 'mathematical structure'. It is then little wonder that mathematicians are sceptical towards the idea of proofs presented by computer calculations (MacKenzie, 1999); the ideology of the professional mathematician emphasizes the conventional operating procedures centred on the concept of the mathematical proof-making. Boundary work is therefore contingent and context-bound, shifting the professional boundaries as new institutions or technologies are changing the field. As emphasized by Strauss, Schatzman, Bucher *et al.* (1964), professions are emerging properties and new social conditions may render certain beliefs and assumptions obsolete. Dent (2003: 108) discusses how traditional professional groups have been increasingly loosened up. First there has been 'an erosion of the professional-managerial divide' where there is no longer of necessity an antagonist relationship between professionals and managers. For instance, professionals are increasingly involved in managerial tasks. Second, in certain sectors, such as accounting and engineering, there has been an increased acceptance for what may be called an 'commercialized professionalism', a more positive attitude towards the need for serving market needs and satisfying customers' and clients' demands. Third, the very notions of 'the professional' and 'professionalism' has been part of what Dent (2003) calls the 'managerial armoury', that is, 'a means of encouraging compliance by extolling the virtue of "professionalism" as a disciplinary logic among employees' (Dent, 2003: 108). That is, various occupational groups are expected to think of themselves as enterprising professionals serving a specific segment of the market, thereby taking on self-monitoring tasks and accepting to be held accountable for their actions.

In summary, the concepts of professions and occupational groups have emerging properties and are the outcome from the thoughtful convergence to professional ideologies and traditions, environmental factors and contingencies, and actual practices. Professional groups also relate to one another in various ways, for instance, in the day-to-day boundary work separating or integrating different occupations or professional groups. Speaking in terms of the construction industry, a large number of occupational and professional groups are involved in the day-to-day work of designing and constructing a building. Ranging from designers, civil engineers and architects to construction workers, plumbers, foremen and site managers, the construction project is populated by a large number of individuals having complementary skills and domains of expertise.

The concept of practice

Professional and occupational teams engage in everyday work-life practices, that is, quite simply, what they do and say in an organizational setting (Gherardi, 2006). However, in the following, the concept of practice is de-familiarized and discussed from an analytical perspective, making the term a tool for investigation into the domain of the everyday work. The concept of practice has been discussed from many theoretical and epistemological perspectives and Schatzki, Knorr Cetina and Savigny (2001) even propose a 'practice turn' in the social sciences. The concept of practice is often associated with a Marxist theoretical framework – Antonio Gramsci (1971), imprisoned by the Italian Fascist regime in the 1920s, spoke of Marx's work as the 'philosophy of practice' to avoid censorship – but has since the 1960s been recognized more broadly in the social and historical sciences, marking an analytical turn away from structuralist and functionalist approaches (Vann and Bowker, 2001: 247). In particular, educational research and research on work have advanced a practice-based view in the social sciences. There are a large number of definitions of 'practice'. Kwinter (2001: 14) speaks of practices as 'particular clusters of action, affectivity, and matter'. Schatzki (2002: 73), dedicating an entire book to the dynamic relationships between the components of what he calls the 'site of the social', speaks of practice as 'a set of doings and sayings'. He continues: 'Practices establish particular arrangements. These arrangements are defined packages of entities, relations, meanings, and positions, whose integrity derives from the organization of practices' (Schatzki, 2002: 101). Acker (2006) advances a rather complex definition of the term:

> Practice, as I use the term, includes production of material 'things', virtual nonmaterial 'things', the material and emotional production of human beings, and the ordinary activities of daily living. Practice is always infused with meaning, and usually informed by thought, although many ordinary activities are guided by tacit knowledge, not consciously invoked.
>
> (Acker, 2006: 46)

Alasdair MacIntyre's *After Virtue* (1981) is one of the most frequently referenced works discussing social practice. Here MacIntyre (1981: 194) distinguishes between *practices* and *institutions*: 'Chess, physics and medicine are practices; chess clubs, laboratories, universities and hospitals are institutions'. Practices are always inextricably entangled with institutions; no practice is capable of surviving for any length of time 'unsustained by institutions': 'Indeed so intimate is the relationship of practices to institutions … that institutions and practices characteristically form a single causal order' (MacIntyre, 1981: 194). Practices and institutions have a recursive relationship: on the one hand, institutions are always the product of practices, on the other it is the institution that enables the practice to be constituted qua legitimate social action (see Giddens, 1984). In addition, as Schatzki (2002) emphasizes, practices are also what constitutes meaning in social systems; meaning is 'established in social practices'. Seen in this view, meaning is not a matter of 'difference, abstract schema, or attributional relativity' (Schatzki, 2002: 58) but is 'a reality laid down in the regimes of activity and intelligibility called "practice"'. Concepts such as practice, institution, and meaning are thus intimately related and Schatzki (2002: 209) even claims that agency is recursively related to practices. Such posthumanist declarations mean that an agent is 'both an arrangement and an effect thereof' – agency is both what enables practices but the agent is also an effect of accumulated practices. Agency is then what structures practices and what is constituted *in situ* and *in actu*. Social practice is then, in Schatzki's (2002) account, a central process for understanding what he calls 'the site of the social', an assemblage comprising practices, material resources, beliefs and norms, and so forth; the site of the social is 'a mesh of practices and orders: a contingently and differentially evolving configuration of organized activities and arrangements' (Schatzki, 2002: xii).

Expressing Schatzki's overtly theoretical elaborations in more mundane terms, everyday work-practice needs to be examined as a composite social practice wherein one must not assume that one component precedes the other. Instead, materiality, practice, cognition and whatever human or social resource is employed are mutually constituted in the very network of practices and mechanisms. Agency is therefore dynamic and emergent rather than confined and already defined once-and-for-all. Schatzki thus provides a fluid and fluxing image of practice that enables new perspectives on everyday work.

Another term that is useful when examining complex social practices is the term 'tinkering' employed by Karin Knorr Cetina (1981) in her study of laboratory work. Drawing on the anthropology of Claude Lévy-Strauss and his term *bricoleur* denoting the handyman capable of effectively using all conceivable resources available, Knorr Cetina argues that laboratory practice emerges not as well-ordered and structured as commonly believed but as a form of tinkering, that is, in Clark and Fujimura's (1992: 11) formulation, 'using what is at hand, making-do, using things for new purposes, patching things together, and so on'. Furthermore, the social science literature provides a series of texts arguing in favour of taking material components into account when examining social practice (Knorr Cetina, 1997; Harré, 2002; Law, 2002; Pels, Hetherington and Vendenberghe, 2002; Mackenzie, 2005). These contributors argue that there is never a purely social domain devoid of technology or artefacts to explore and describe, but all social action is bound up and entangled with material resources. Orlikowski (2007: 1437), speaking from an organization theory perspective, is arguing that materiality is 'integral to organizing' and that the social and the material are 'constitutively entangled in everyday life'. In such a perspective, Orlikowski argues, neither the social nor the material is privileged but they are 'inextricably related' – 'there is no social and that is not also material, and no material that is also social' (Orlikowski, 2007: 1437). The concept of tinkering is affirmative of such a perspective, recognizing that material and social resources are always co-dependent. In addition, tinkering is the capacity of 'making things work', to pull together all the resources needed to accomplish whatever is desired (Timmermans and Berg, 1997: 293). While practice is a term that serves the purpose of denoting a social practice that is embedded in institutions and the broader environment, tinkering is emphasizing the local milieu or even the specific setting where the individual is operating and the importance of the materiality engaged in the work. Tinkering is therefore complementing the concept of practice in terms of being even more down to the earth and underlining the use of actual available resources in one's work.

Yet another term pertaining to day-to-day work practices is the term 'action nets' proposed by Czarniawska (1997). For Czarniawska, it is not particularly meaningful to focus on individual co-workers' activities in an organization: '[o]rganizations are not people at all (neither aggregates nor collectives of super-persons); they are *nets of collective actions* undertaken in an effort to shape the world and human lives', Czarniawska (1997: 41) argues. Instead, it is the collective net of actions, constituted by practices, materials and intangible resources such as ideologies that should be examined. Consistent with what has been called 'garbage-can decision-making' processes where individuals come and go during a decision process (Cohen, March and Olsen, 1972; March and Olsen, 1976), the concept of action nets is formulated to tolerate such exits and entrances on the part of organizational members. Czarniawska explicates the idea:

> Action nets are neither people nor groups; they may be large (across several organization fields) or small (a project); the focus of analysis can be a combination or collection of such nets (an organization field). It is from the action net that we deduce which actors are involved, not the other way around. This means, for example, that the net will never continue to exist even when the actors are exchanged for others, or the original actors change their identity (they may become machines), although it always means the change in the character of the net as well; that the changing net may press for a change in the identity of the actors ... that the actors may be of fixed status (humans and nonhumans; Latour, 1992a, 1992b) – a fact we would miss if we looked exclusively at human actors and their interactions.
>
> (Czarniawska, 1997: 179)

An action net is then an 'anti-essentialist' view of organizing while the solid effects of organizing are recognized (Czarniawska, 2004: 780). In addition, the concept is supposed to minimize what is 'taken for granted' prior to any empirical investigation:

> a 'standard analysis' begins with actors or organizations but an action net approach permits us to notice that these are the products rather than the sources of the organizing – taking place within, enabled by and constitutive of an action net. Identities are produced by and in an action net, not vice versa.
>
> (Czarniawska, 2004: 780)

Furthermore, an action net approach does not assume that collective action is structured within one single organization; instead, an action net may involve a 'great variety' of organizations or organized groups of people (Lindberg and Czarniawska, 2006: 293). Seen in this view, an action net is a useful perspective when studying practice because it does not assume that practices are predefined or already given. Instead, practices are what emerge in action nets, which in turn recursively provide a shared ground for further action. The concept of action net thus shares a non-essentialist view of practice with Schatzki's concept of 'the site of the social'. Both these theoretical frameworks regard social actions as what constitutes social reality and action nets, respectively, but without reifying social practices as being once and for all given.

In summary, the concept of practice is useful when examining knowledge-management activities because it recognizes the local and contingent social action of individuals and groups of individuals. Just like the concepts of professional and occupational groups is helpful for understanding how individuals are taking on identities and adhering, more or less consciously, to professional standards and ideologies, the concept of practice is an analytical tool that helps bridge the particular and local and the universal, the idiosyncratic and the collective. Using Timmermans and Berg's (1997) term 'local universality' underlining that 'universality always rests on real-time work, and emerges from localized processes of negotiations and pre-existing institutional, infrastructural, and material relations', that is, universality is what is constituted in the bits and pieces of everyday life rather than being imposed from above as some kind of transcendental category, one may argue that practices are always both local and universal; they draw on shared institutions but they are always carried out individually and locally.

The concepts of knowledge and knowledge management

The establishment of a knowledge-management perspective in organizations

The concept of knowledge is, needless to say, one of the most complicated and multifaceted concepts in the contemporary social science vocabulary. It is beyond the scope of this chapter to provide full coverage of the philosophical, sociological and economical implications from the term (for an overview, see Styhre, 2003) but some of the central intricacies of the term will be highlighted. In the philosophical canon, the concept of knowledge has been part of the *philosophia perennis*, at least since Plato addressed the concept of knowledge in dialogues such as *Meno* and *Thaeatetus* in the third century BC. The sub-discipline of epistemology (derived from the Greek word *episteme*, 'knowledge') is actually dedicated to issues regarding the nature and limit of knowledge. In contemporary sociology, the term 'knowledge' has been examined as one of the most significant production factors of our time (e.g. Touraine, 1971; Bell, 1973), thereby restructuring the economy and social structure when predominant regimes of knowledge (e.g. knowledge in agriculture or manufacturing) is displaced by new forms of knowledge (in e.g. the computer sciences or in design work). For instance, in the early seventies, the French sociologist Alain Touraine (1971: 51) declared, contrary to Marxists doctrines, that, rather than drawing on inherited financial resources and cultural capital, '[t]he new dominant class is defined by knowledge and a certain level of education'. In a similar vein, Bell (1973) announced the beginning of 'post-industrial society' characterized by its reliance in specific forms of knowledge. Even though the field of organization theory and management studies has from the outset (in e.g. the quickly growing body of literature on scientific management and systematic management – for instance, Henry Gantt (1919: 161) one of Frederick W. Taylor's most well-known disciples, carefully distinguished between 'expert' and 'standard'knowledge in accordance with Taylor's principles) addressed the use of various cognitive resources, such as skills and know-how in companies and organizations, it was not until the early 1960s that organization theorists more explicitly addressed the concept of knowledge. For instance, in his 1964 book *Modern Organizations,* Etzioni (1964: 75) claimed that '[m]ost knowledge is created in organizations and passed from generation to generation – i.e. preserved – by organizations'. Two years earlier, Fritz Machlup (1962) published what is arguably the first knowledge management book, *The Production and Distribution of Knowledge in the United States.* Here Machlup sketches how knowledge is increasingly influencing organizations and the broader society. Machlup (1962: 21–2) distinguishes between five categories of knowledge: (1) *practical knowledge* in which he includes 'professional knowledge', 'business knowledge', 'workman's knowledge, 'political knowledge' and 'household knowledge'; (2) *intellectual knowledge,* for instance scientific skills and capabilities; (3) *small-talk* and *pastime knowledge,* that which is used in everyday life to support social interaction and relationships; (4) *spiritual knowledge*; and (5) *'unwanted knowledge',* knowledge one possesses but does not really care for or would prefer not to be aware of. Etzioni (1964) distinguishes between three types of organization in terms of their use of knowledge:

- *Professional organizations* that hire at least 50 per cent professionals that are capable of producing, applying, preserving and communicating knowledge.
- *Service organizations* where professionals are subordinated to administrators, and
- *Non-professional organizations,* such as manufacturing firms and the 'military establishements'.

(Etzioni, 1964: 77)

After the foundational works of Etzioni (1964) and Machlup (1962) were published in the early 1960s, organization theory researchers were preoccupied with other concerns and the issue of knowledge was very much abandoned. However, in the middle of the 1990s, a few seminal works were published addressing knowledge as a significant organizational resource when explaining the differences in performance between organizations (Leonard-Barton, 1995; Nonaka and Takeushi, 1995). In 1996, *Strategic Management Journal* published a special issue of the 'knowledge-based view of the firm' edited by Robert Grant and J.-C. Spender (Spender and Grant, 1996), examining the corporation as a repository of knowledge. The 'knowledge-based view of the firm' was here essentially conceptualized in dialogue with the ongoing debate in the journal on the relationship between the resource-based view (RBV) of the firm (represented by, for instance, Barney, 1991, 2001), and traditional industrial organization economics perspectives on strategy (Bain, 1968; Conner, 1991). Rather than speaking of competitive advantage in terms of positioning in the market, RBV theorists suggested that one should examine the specific resources of the firm to explain 'extranormal profits' (or 'rent' as the strategic management theorists prefer to speak of) generated. In this framework, knowledge is conceived of as the single most important resource and therefore it needs to be theorized as such. Beginning in the mid-1990s, knowledge management became one of the new sub-disciplines in organization theory, engendering new conferences, journals, professional associations, and professional consulting services and offers. Today, some 15 years after the newly awakened interest in the concept of knowledge, this theoretical perspective has been thoroughly established. The publication of no less than two 'handbooks' on knowledge management testifies to the gradual institutionalization of a knowledge-management perspective on organizations (Dierkes, Berthon, Child and Nonaka, 2001; Easterby-Smith and Lyles, 2003).

Taxonomies of knowledge

In their review of the knowledge management literature, Amin and Cohendet (2004: 5–6) identify three complementary perspectives on knowledge. First, they speak of a strategic-management approach where concepts such as *core competencies, competitive advantage,* and *intellectual capital* are used to examine organizations and firms as repositories of knowledge. Second, they point at an *evolutionary-economics approach* represented by, for instance, Nelson and Winter's (1982) influential work, where the knowledge of the firm is embodied in its routines and standard operating procedures. Third, and most relevant for the forthcoming discussions in this book, a *social-anthropology-of-learning approach* is named. In this last approach, knowledge is not what is laid down in organizational structures and routines or otherwise manifested in organizational activities, but is instead what is actively produced in collective action in organizations. Amin and Cohendet (2004) are not the only authors seeking to sort out categories of knowledge into taxonomies. In fact, quite a large number of papers and books contribute to a structuring of the literature into, if not mutually excluding, at least complementary categories. For instance,

Sackmann (1992: 141) speaks of *dictionary knowledge* addressing 'what?' questions, *directory knowledge* addressing 'how?' questions, *recipe knowledge* addressing 'should?' questions, and *axiomatic knowledge,* addressing 'why?' questions. Knowledge here ranges from descriptions of actual or perceived conditions to declarative statements of cause-and-effect relationships and judgements. One common line of demarcation between types of knowledge is that between the propositional knowledge of the expert, and the local and performative know-how of everyday life. For instance, Yanow (2004) sketches these differences in Table 1.1.

A similar distinction is made by Empson (2001), who argues that one may either regard knowledge as assets ready to be exploited or as a process where joint collaborations between actors constitute what counts as knowledge. Empson's two perspectives are summarized in Table 1.2.

The distinction advocated by Empson (2001) is representative of the separation between propositional knowledge ('know-that') and performative knowledge ('know-how') advanced by Gilbert Ryle (1949). Propositional knowledge is referring to actual condition (e.g. 'I know that in Paris, most people speak French'), while performative knowledge is the capacity to actually perform a task (e.g. 'I do not speak French'). Despite all its richness, the English language does not have adequate terms for distinguishing between these two forms of knowledge. In German for instance, it is common to distinguish between *können* ('know-how') and *wissen* ('know-that') and similar linguistic resources are available in classic Greek, French and the Scandinavian languages. Gourlay (2006) provides an overview of how the distinction between the two forms of knowledge has been influencing a range of disciplines (Table 1.3).

In the following, the local and processual view of knowledge, the knowledge that emphasizes know-how and the performative function of knowledge, will be discussed in greater detail. In doing so, we follow Tsoukas and Mylonopoulos (2004: S3) in their insistence on 'not taking knowledge for granted' or 'assuming it has already a particular form and content', but to actively engage in critically examining the nature and function of knowledge.

Table 1.1 Two types of knowledge (Adapted from Yanow, 2004: S12)

'Expert knowledge'	'Local knowledge'
Theory-based	Practice-based
Abstracted, generalized	Context-specific
Scientifically constructed	Interactively-derived
Academy-based	Lived experience-based
Technical-professional	Practical reasoning
Explicit	Tacit
Scholarly	Everyday

Table 1.2 Alternative perspectives on knowledge in organizations (Adapted from Empson, 2001: 813)

	Knowledge as an asset	*Knowledge as a process*
Purpose of research	Normative. To identify valuable knowledge and to develop effective mechanisms for managing that knowledge within organizations	Descriptive. To understand how knowledge is created, articulated, disseminated, and legitimized within organizations
Disciplinary foundation	Economics	Sociology
Underlying paradigm	Functionalist	Interpretive
Epistemological assumptions	Knowledge as an objectively definable commodity	Knowledge as a social construct
Models of knowledge transmission	Exchange of knowledge among individuals is governed by an implicit internal market within organizations	Knowledge is disseminated and legitimated within organizations through an ongoing process of interaction among individuals
Main level of analysis	Organization and its knowledge base	Individuals in social contexts

Table 1.3 Knowledge types and names (Adapted from Gourlay, 2006: 1426)

Discipline	*Knowledge-how*	*Knowledge-that*
Philosophy	Knowledge-how; procedural knowledge; abilities	Knowledge-that; propositional knowledge
Philosophy (Polanyi)	Tacit knowledge	Explicit knowledge
Psychology	Implicit knowledge; tacit abilities; skills	Explicit knowledge; declarative knowledge
Artificial intelligence	Procedural knowledge	Declarative knowledge
Neuroscience	Covert knowledge	Overt knowledge
Management studies; education	Tacit knowledge	Explicit knowledge
Information Technology studies	Knowledge as process	Knowledge as object
Knowledge management	Know-how	Know-what
Sociology of science	Tacit; encultured (forms of life)	Explicit/symbolic

The social constitution of knowledge

When following what Amin and Cohendet (2004) call the *social-anthropology-of-learning approach* to knowledge management, it is important to recognize the social constitution of knowledge. A number of contributions to this literature emphasize the social embedding of knowledge. Sole and Edmondson (2002: S20) speak of what they refer to as 'situated knowledge' as what is embedded in the work practices of a particular organization or work place. Knowledge is here what derives from intimate familiarity with how things work in practice and how to muddle through all the challenges of emerging in everyday work-life. Sole and Edmondson's (2002) idea of a 'situated knowledge' is shared by Gherardi and Nicolini (2001: 44) who emphasize that no knowledge is 'universal or supreme', but always produced within social, historical and linguistic relations that are grounded in 'specific forms of conflict and the division of labor'. For Tsoukas and Vladimirou (2001: 990), knowledge management – the active manipulation and control of knowledge resources in an organization – is therefore a dynamic process of 'turning an unreflected practice into a reflective one by elucidating the rules guiding the activities of the practice, by helping give a particular shape to collective understandings, and by facilitating the emergence of heuristic knowledge'. 'Knowledge is based', Bourdieu (2004: 72) argues, 'not on the subjective self-evidence of an isolated individual but on collective experience, regulated by norms of communication and argumentation'. Managing knowledge is then not solely to handle information but actively to sustain and strengthen social practices; knowledge management is, in short, a matter of *socialization* (Tsoukas and Vladimirou, 2001: 991). The perspectives on knowledge management pursued by Sole and Edmondson (2002), Gherardi and Nicolini (2001) and Tsoukas and Vladimirou (2001) emphasize that all knowledge is 'socially overdetermined' – it derives from many sources and emerges in diverse forms. Seen in this view, one may argue that it is not meaningful to conceive of knowledge in terms of linearity and cause–effect schemes. Instead, one may think of knowledge in more dynamic terms, as what Castoriadis (1997) calls the 'co-production of knowledge'. When speaking of co-production, we 'cannot truly separate out what "comes from" the subject and what "comes from" the object' (Castoriadis, 1997: 345). Castoridis suggests the term 'the principle of undecidability of origin' to denote this dynamic relationship between the object and the subject. Studies of the laboratory sciences (e.g. Fujimura, 1996; Rheinberger, 1997) suggest that scientific knowledge emerges in the dynamic inter-relationship between theoretical frameworks, laboratory equipment, scientific specimens, such as laboratory animals, and laboratory practices. In the laboratory milieu, knowledge is what is both constituted and applied in what Pickering (1995) calls 'the mangle of practice'. Patriotta (2003) acknowledges this social embeddedness of knowledge and consequently speaks of 'knowledge-in-the-making', knowledge that is in a state of becoming. In addition, knowledge is here always 'controversial, ephemeral, and experimental'. In common-sense thinking, knowledge is generally conceived of as coming and circulating in 'canned packages' (Lanzara and Patriotta, 2001: 944). Knowledge is then taken for granted and rendered a certain degree of 'facticity', a 'matter-of-factness' that more or less ignores or 'brackets' the interactive and contentious nature of knowledge. Contrary to this 'ready-made' view of knowledge, knowledge-in-the-making recognizes the tensions internal to all claims to knowledge and emphasizes the social components in all forms of knowledge. Lanzara and Patriotta (2001) here introduces the concept of *assemblage* (DeLanda, 2006) to further underline the heterogeneous and social embeddedness of knowledge:

> Rather than a discrete commodity, organizational knowledge could be better pictured as an 'assemblage' subject to continuous transformations and reconfigurations. It is an assemblage precisely because it is the outcome of controversy and *bricolage*, resilient as a whole but subject to local disputed, experiments, and resembling ... An assemblage is neither a unity not a totality, but a multiplicity, a collection of heterogeneous materials that are mutually but loosely interrelated. In other words, the notion stresses the importance of relations over the elementary parts, i.e. what goes on 'between' the part (Cooper, 1998, p. 112). In this regard, what makes knowledge distinctive is not the discrete collection of commodities, but the nature of the assemblage and, we should add, the making of the assemblage in time. An assemblage is an evolving artifact and it is unique because it springs out of unique history. In summary, the notion of assemblage emphasizes the pasted-up, path-dependent nature of knowledge systems and reinforces the definition of knowledge as a phenomenon in the making, which eventually make sense in the retrospect.
>
> (Lanzara and Patriotta, 2001: 964)

The various elements of a body of knowledge is here loosely coupled frameworks, constituted by, for instance, theories, experimental practices, everyday practices, common-sense beliefs, statements and declarations issued by authorities and regulatory bodies, and so forth. There are many rationales underlying knowledge claims, ranging from traditional beliefs and mythologies to advanced technoscientific research. The concept of knowledge draws on such heterogeneous resources and is consequently what is always multiple, always 'in-the-making'.

In the *social-anthropology-of-learning approach* to knowledge management outlined by Amin and Cohendet (2004), knowledge is a social accomplishment that needs to be examined as what integrates many actors, artefacts and other relevant resources. Knowledge is here the outcome from social practices in professional or occupational groups, but is simultaneously what structure and organize such groups. Knowledge is what is inherently social and needs to be examined as such.

Knowledge workers, knowledge-intensive firms, and so forth

In the knowledge management discourse – here defined as a 'regulated system of statement' (Hassard and Parker, 1993: 63) on a specific topic having implications for social practice – a number of subject-positions and institutions have been established. First, there is a new emerging class of workers which has been referred to with terms such as *knowledge workers* (Schultze, 2000), *the creative class* (Florida, 2002), *the new dominant class* (Touraine, 1971) or *symbol analysts* (Reich, 1991). Schultze (2000: 5) argues that knowledge workers form a special class of white-collar workers, including 'professionals, consultants, technicians, intellectuals, and managers' and shares a number of characteristics:

- It produces and reproduces information and knowledge.
- Unlike physical blue-collar work, knowledge work is cerebral ... and involves the manipulation of abstractions and symbols that both *represent the world* and are objects *in the world.*
- Unlike *service work*, which is frequently scripted ... knowledge work defies routinization and requires the use of creativity in order to produce idiosyncratic, esoteric knowledge.
- It requires formal education, i.e. abstract, technical and theoretical knowledge.

(Schultze, 2000: 5)

Heervagen *et al.* (2004: 511) claim that 'knowledge work tasks' include planning, analyzing, interpreting, developing, and creating products and services using information, data or ideas as the raw materials. In the analysis of Heervagen *et al.* (2004: 511), the day-to-day work of knowledge workers is characterized by a series of short periods of concentrated, intensive work, combined with brief interactions with colleagues:

- Workers have small blocks of uninterrupted time, punctuated by frequent, brief conversations.
- At any given time, only a proportion of tasks are worked on, with multiple tasks being in a state of suspension.
- Task switching is common and results, in large parts, from interruptions to on-going work.
- People spend most of their interaction face-to-face.
- Most face-to-face interactions at work are opportunistic rather than planned.

(Heervagen *et al.* (2004: 511–12)

Being of a specific social class, knowledge workers develop their own ideologies and standards for what qualifies as legitimate work in the domain of expertise. The class of knowledge workers has its own qualifying criteria for being a member of the class and the entrance into the community is monitored by what Pettigrew (1973), speaking about technology acquisition and not knowledge work, called 'gate-keepers'. As sociologists subscribing to class as a central analytical category (e.g. Dahrendorf, 1959) suggest, class is a category that always assumes a certain degree of conflict and struggle over resources at the structural level, and the concept of knowledge workers as social class is no exception; knowledge work is firmly rooted in the line of demarcation between knowledge and pseudo- or 'non-knowledge', or knowledge belonging to other domains of expertise. The institutions and affiliations provided by the knowledge worker class help to signal what members of the community are regarded as legitimate spokespersons on specific topics. As Alvesson (2001: 872) points out, 'as a layman, one "knows" that a person is knowledgeable because credible institutions have declared that such is the case. An expert belongs to a community of experts: authorization and membership of associations are often the criteria for expertise'. Taking on the role of the expert is therefore based on the illocutionary speech act (Austin, 1962), wherein one declares one's affiliation with a specific community (e.g. a lawyers' association) or institution (e.g. a university). As an analytical category, the concept of knowledge worker has its merits, bridging the concepts of professions and occupational groups and practices in individual organizations or networks of organizations.

Another frequently used term in the knowledge management discourse is the concept of *knowledge-intensive firms* (Starbuck, 1992). A knowledge-intensive firm is, quite simply, an organization hiring primarily knowledge workers; it is similar to what Etzioni (1964) called the 'professional firm', but may not hire only 'professionals' but any group of individuals specializing in a domain of expertise. Alvesson (2000: 1101) argues that typical examples of knowledge-intensive firms include 'law and accounting firms, management, engineering and computer consultancy companies, advertising companies, R&D units, and high tech companies'. Since knowledge, the primary 'production factor' by definition in any knowledge-intensive firm, tends to be tacit and residing in cognitive, embodied and emotional human capacities, this knowledge is inherently complicated to control and manage. Knowledge-intensive firms are in short 'people-dependent organizations' (Robertson and Swan, 2004: 124). Therefore, it is often pointed out that knowledge-intensive work is not managed through conventional means but new forms of control need to be employed. Kärreman and Alvesson (2004: 151) emphasize that knowledge-intensive firms are using what they call 'cultural-ideological modes of control' to manage their operations and day-to-day work. 'Cultural-ideological modes of control' are controls that seek actively to promote certain beliefs, norms, identities and subject-positions (e.g. as programmer, real estate agent, or surgeon) to structure and control the day-to-day work in the organization. Robertson and Swan (2003: 836) use the term 'normative control' to denote the new regime of control serving to 'self-discipline' co-workers. One such 'genre' within the domain of 'normative control' is the idea of 'entrepreneurialism' or the 'enterprising self', that is, the integration of conventional managerial roles into one's work description. In the entrepreneurialism genre, it is the co-worker, not management, which is held accountable for the performance. Robertson and Swan (2004: 128) report that 90 per cent of the firms in a study relied on 'either cultural or professional forms of control' rather than 'bureaucratic forms of control' (denoting traditional forms of performance or work process control). The everyday work-life of the knowledge worker is thus characterized not only by the demand continuously to update one's skills and capabilities but also to monitor one's own performance and report to management. Knowledge workers are therefore *de facto* operating as semi-autonomous entrepreneurs, in many cases catering for their own stock of clients or customers.

Some further remarks on the nature of knowledge

In addition to being inherently social and brought into action in knowledge-intensive firms employing new means of normative cultural-ideological control, knowledge is a concept with significant epistemological ramifications. As has been pointed out by numerous philosophers and social theorists, knowledge is never a self-enclosed category but is always fundamentally open to influences and diverse human faculties. For instance, Alfred North Whitehead (1967: 4) regards the idea of 'mere knowledge', knowledge enclosed unto itself, as a 'high abstraction'. Instead, knowledge is always 'accompanied with accessories of emotion and purpose', Whitehead claims. The tendency to reduce knowledge to mere cognitive or rational thinking is a fallacy that is addressed in many texts looking at the nature of knowledge. For instance, Salzer-Mörling (2002: 117) underlines the importance of the five human senses in any knowledge work:

> [T]he reductionism of organization to purely cognitive scripts not only preclude the sensory faculties of hearing, smelling, sight, touch and taste from organizational life, but also neglects the physical manifestations and extension of knowledge. Organizing is not a purely intangible process, rather it is made up of 'heterogeneous materials', involving all the senses.

Another important trait of knowledge-intensive work pertaining to work in the construction industry is that knowledge resides in practices and routines and therefore 'applying is preserving' (Starbuck, 1992). Knowledge cannot always be 'stored' but people 'must relate it to their current problems and activities' (Starbuck, 1992: 772). Starbuck (1992) continues: 'Effective preserving looks much like applying. As time passes and social and technological changes add up, the needed translation grows larger, and applying knowledge comes to look more like creating knowledge'. Knowledge is what is in need of constantly being updated or else it vanishes away. Know-how is therefore the 'agent's knowledge' (Tsoukas and Mylonopoulos, 2004: 5), the knowledge of actual practices and not the propositional knowledge of the 'spectator'. The agent's knowledge is here involving and drawing on a large variety of human faculties and skills including emotionality and embodiment. While the early stages of knowledge management often addressed the management of knowledge as a form of systematic handling of information in organizations (Scarbrough and Swan, 2001; and see Table 1.4), today, there is much more elaborated and theoretically sophisticated literature addressing the concept of knowledge and its use in organizations. Knowledge is no longer conceived of as a solely cognitive construct, but concepts such as emotionality, embodiment and materiality have been fruitfully accommodated within the discourse on knowledge management.

In summary, knowledge management has emerged as a body of literature conceiving of firms and organizations as being fundamentally embedded in the active and thoughtful use of their knowledge resources and skills. It is an analytical perspective which helps in the understanding of advanced operations and activities that involve specialized skills and domains of expertise.

Knowledge management in the construction industry

Previous studies of knowledge management in the construction industry

The literature on knowledge management in the construction industry is rather limited, both in Sweden and internationally. Carrillo, Robinson, Al-Ghassani and Anumba (2004) and Robinson, Carrillo, Anumba and Al-Ghassani (2005) (belonging to the same research team) study the use

Table 1.4 1998 knowledge management thematic categories (Adapted from Scarborough and Swan, 2001: 8)

Thematic category	Number	per cent
Information technology	73	40
Information systems	51	28
Strategic management	35	19
Human resources	9	5
Consultancies	8	4
Others: libraries, academic, accounting, marketing	8	4

of knowledge management in the UK construction industry. They found that a majority of the companies actively used knowledge-management practices and in 61.9 per cent of the cases a knowledge manager had been assigned the role of being responsible for the firm's knowledge management activities (Carrillo, Robinson, Al-Ghassani and Anumba, 2004: 51). These authors show that 63 per cent of the responding firms in a survey regarded their knowledge-management work as being 'an *ad hoc* process'; furthermore, they suggest that 'the lack of standard work processes' is regarded as the single most important impediment towards the use of knowledge management practices (Carrillo, Robinson, Al-Ghassani and Anumba, 2004: 50ff). However, Carrillo, Robinson, Al-Ghassani and Anumba (2004) and Robinson, Carrillo, Anumba and Al-Ghassani (2005) do not examine in great detail what activities the construction firms undertake under the label 'knowledge management'. A similar argument is provided in Kamara, Augenbroe and Carrillo (2002), saying that a variety of firms in the construction industry do in fact use knowledge management practices but they fail to account for how the daily routines and activities are carried out. In one of the few edited books on knowledge management in the construction industry, also examining the UK, a similar perspective is taken (Kazi, 2005). Here, the industry is reviewed in terms of *what* knowledge-management practices are used rather than *how* such practices are performed. The edited volume of Kazi (2005) does then provide only a rather limited insight into how knowledge management is actually influencing the day-to-day work. Bresnen, Goussevskaia and Swan (2004) emphasize the complexity and the large number of relations between different actors in construction projects and point at the problems associated with the systematic creation, sharing and distribution of knowledge in such an organizational setting. The paper examines how the knowledge-management tool 'The Dashboard' is implemented and received in three regions in a construction company. In their paper, Bresnen, Goussevskaia and Swan (2004) thus seek, in contrast to most other texts, to make the *practice* of knowledge management a source of investigation.

Drejer and Vinding (2002) present survey-based research showing that firms that have established what they call 'knowledge anchoring processes' such as systematic project evaluation, examination of finished buildings, and mechanisms for distributing and sharing know-how are more likely to be more innovative. Even more innovative were construction companies participating in partnering with other construction firms, that is, a form of joint venture where the two firms share their insights and experiences throughout the focal research project. Drejer and Vinding (2002) articulate some managerial implications from their work:

> The lesson for managers is that knowledge-anchoring mechanisms and partnering may help reduce the shortcomings of project-based organizations as regards capturing, sharing, and diffusing knowledge and

learning across projects, and instead become more knowledge-driven. The use of post-projects review and systematic evaluations and diffusion of experiences means that managers may have less difficulty in combining strategies of short-term task performance with long-term learning and knowledge accumulation.

(Drejer and Vinding, 2006: 928)

Innovative firms are in other words firms that have established clear mechanisms and procedures for developing, sharing and distributing knowledge. Such 'knowledge-anchoring mechanisms' may help firms overcome the inertia imposed by the project and network form predominating in the industry. Carrillo's (2004) study of the gas and oil industries – industries she claims share that many characteristics with the construction industry – is providing similar results. Carrillo (2004: 640) recommends what she calls 'people-centred techniques' for the sharing of tacit knowledge and information technology (IT) tools for the sharing of explicit knowledge. She also found that all knowledge is embedded in social relations and the social capital of the firm and consequently peer review has a 'more sustainable impact' than financial rewards (Carrillo, 2004: 641).

Rooke and Clark (2005) present an ethnographic study of how safety knowledge is developed and shared at a construction site in the UK. They notice that knowledge is passed from individual to individual, but that it is unlikely to be 'adequately recorded' (Rooke and Clark, 2005: 562). Construction workers have in general three ways of learning: by watching more experienced co-workers, by trying things out, or by direct instructions (Rooke and Clark, 2005: 566). There is, in other words, a rather practical approach to learning; instructions accompanied by observations and eventually actual activities make up a 'learning cycle' constituting the 'experiential knowledge' of the construction workers. Rooke and Clark (2005) also found that the experiential knowledge valued by the construction workers at times clashed with the engineers' scientifically based, 'classroom-taught' knowledge. As a consequence, 'conceptions of knowledge and authority operate at the site level to impede attempts at improving safety performance', Rooke and Clark (2005: 568) conclude. Engineers' attempts to enhance safety performance were rejected on the basis of its inconsistency with personal experiences and know-how, embedded in actual work rather than formal education.

The literature concerning the Swedish construction industry is also rather limited. Sverlinger (2000) made a quantitative study (a survey) of knowledge transfer within four large Swedish technical consultant firms. Managers confirmed the existence of strategies for knowledge management, but these remained largely unknown to employees, and responsibility for knowledge-management efforts was unclear. Sverlinger (2000) also identified six enablers for knowledge transfer: organizational structure, communication and monitoring of strategy, process, culture, systems for training and learning, and technology. In summary, the literature on knowledge management in the construction industry is rather sketchy, saying that there are examples of the use of knowledge-management procedures but reporting little detailed accounts of how knowledge management is practised in day-to-day work-life. The literature primarily provides an inventory analysis, listing things construction companies do or do not do. Moreover, the research points out that many companies use the concept of knowledge management as synonymous with the management of information (Robinson, Carrillo, Anumba and Al-Ghassani, 2005: 437). Therefore, the literature emphasizes the storing and sharing of information rather than the active promotion of knowledge inherent to practices.

Some characteristics of the construction industry

Construction work as an oral culture

One of the longstanding discussions in social theory is that of the importance of writing in relation to verbal communication. In this section, these two forms of communication will be contrasted against one another. In real-life settings, a strict dichotomous view would be simplistic and misplaced, but in a theoretical analysis, the difference between written documents and verbal communication represents two complementary epistemologies and forms of knowledge-sharing. Following this binary model, in the following, on the one hand, we will talk about the perspective that stresses the importance of inscriptions and written documents in the development of social institutions and processes of organizing and, on the other hand, we will explore the perspective that recognizes the verbal, non-written communication between individuals in organizations.

Refusing to adhere to the distinction between 'primitive' and 'civilized' people, the French anthropologist Claude Lévi-Strauss (1992) talks of people with and without writing:

> There are people with, or without writing; the former able to store up their past achievements and to move with ever-increasing rapidity towards the goals they have set for themselves, whereas the latter, being capable of remembering the past beyond the narrow margin of individual memory, seem bound to remain imprisoned in a fluctuating history which will always lack both a beginning and any lasting awareness of an aim
>
> (Lévi-Strauss, 1992: 298)

For Lévi-Strauss, people without writing are necessarily more inclined towards myths and short-term thinking because they are unable to store information (Burns, 1989). Learning is here primarily based on the verbal interaction between individuals, and learning is also dependent on embodied practices where showing and telling are inextricably entangled and an integral component of the learning process. People who do have the ability to write are able to develop more advanced institutions and more abstract mechanisms that do not make sense to illiterate people (see e.g. McLuhan, 1962; Ong, 1982). For instance, the anthropologist Jack Goody (1986) has examined the importance of writing in the development of the modern bureaucratized state administration. For Goody, writing is of great importance but there are still examples of relatively advanced social arrangement in illiterate societies. Goody (1986: 91) writes: '[W]riting is critical in the development of bureaucratic states, even though relatively complex forms of government are possible without it'. He continues: 'Writing was not essential to the development of the state but of a certain type of state, the bureaucratic state' (Goody, 1986: 92). Even though writing is not the only practice that serves as the basis for what Goody calls the bureaucratic state, it certainly enables a more abstract form of thinking; for instance, in terms of separating 'person' from 'office', a most modern invention and a trait of the modern society (Bendix, 1971; Kallinikos, 2003). Goody (1986) writes:

> While some features of the Weberian concept of bureaucracy ... are certainly present in oral societies ... the absence of writing inevitably places limits on the efficacy of government (especially as regards the storage of information) as well as of business firms, churches and other large-scale organizations. As Weber pointed out, one major feature of such administrative organs is the ability to separate 'person' from 'office', people from corporation, and to establish 'universalistic' as against 'particularistic' relationships.
>
> (Goody, 1986: 92)

In Goody's terms, the process of organizing is not exclusively dependent on the practice of writing, but it certainly enables more advanced forms of organization. In addition, the emergence of stable and predictable social institutions, the mark of distinction of modernity, which support and enable social organization and the differentiation of such organization, are also dependent on practices of writing. Speaking of organizations such as firms and corporations, Callon (2002) examines the practices of writing in organizing. Callon (2002: 191) argues: 'Without tools for collecting, constructing, processing, and calculating information, agents would be unable to plan, decide or control. In short, organized action would be impossible'. Callon claims that what he refers to as *writing practices* enables maintaining and recognizing the complexity of the systems, which the organization must manage while at the same time controlling the system: '[T]hey [writing practices] make the complexity of systems of action manageable and controllable without eliminating it' (Callon, 2002: 193). As a consequence, writing practices are, in his view, of great importance for aligning 'doing' and 'knowing', of bringing formal know-how and practices together: '[W]riting devices lie at the heart of the organization in action and ... without them the organization would not exist, as it does, in a location between knowing and doing' (Callon, 2002: 212).

Proponents of the practices of writing offer a number of compelling arguments; writing does in fact have a significant influence on the differentiation of society and the emergence of advanced forms of organization. Proponents of verbal communication as a key mechanism in organizational communication, however, do not refuse to recognize practices of writing. Instead they argue that practices of writing are capable of offering scripts that aim at capturing practices but merely serve as general guidelines or scripts determining the most generic qualities of informed action. Lynch (2002) does, for instance, claim that laboratory work in practice in microbiology is always stretching beyond the written protocols. As a consequence, scientists are participating in a continuous exchange of recommendations, proposals and rules of thumb for making the laboratory work come out as intended. Here, the laboratory is by no means a closed system or a mechanical system but must always be attended to in every instant. For Lynch, it is therefore problematic to think that the protocol is capable of capturing the full complexity of laboratory practices. That would be, with Bateson's (1979: 30) formulation, to 'confuse the map with the territory'. Therefore, the written material used in practice is always generic sketches of actual practices and undertakings. As a consequence, practices are dependent upon verbal and symbolic communication.

Several writers examine organizations as verbal communities. Boden (1994) offers a compelling argument in favour of a view of organization as being essentially a matter of talk and verbal communication. Drawing on symbolic interactionist sociology (e.g. Cooley, 1902; Mead, 1934) and the ethnomethodology of Harold Garfinkel (1967), Boden argues that it is verbal exchanges that make an organization work properly. Boden (1994: 15) writes: 'Through their timing, placing, pacing, and patterning of verbal interaction, organization members actually constitute the organization as a real and practical place'. In this perspective, organization is not as much a matter of physical resources as it is dependent on the circulation and production of information. Furthermore, this information is never wholly transparent for the members of the organization but becomes meaningful in the very process of talk. Boden argues:

> The many coalitions of any organization or group of organizations ... depend not so much on physical resources as on information. The information may be related to material resources but it is usually far more diffuse and multivalent. Much 'news' in organizations are *soft* – incomplete, fuzzy, and flexible – and it is the *blur* that needs

to be classified, refocused, and occasionally completely realigned through talk.

(Boden, 1994: 152)

Following Weick (1979), talking *makes sense* in organizations. In a similar perspective, Donnellon (1996) has examined talk in teams and Leidner (1993) participated in a training programme of sales workers wherein talking was a key competence in becoming successful in the profession. Alvesson (1994) has suggested that the identity of individuals in organizations – this is especially pronounced in professional organizations – is affected by the various forms of talking in the company. Fine (1996) presents a study of the talk and rhetoric of restaurant-kitchen workers. Orr's (1996) much-cited study of copy-machine technicians is another example of how verbal interaction enables a community of practice to share know-how and information, and make sense out of ambiguous events and occurrences. In addition to the talking directly associated with organizational mechanisms such as communication, identity and performance, Kurland and Pelled (2000) and Michelson and Mouly (2000) have examined the 'small talk' of the organization as one significant form of information-sharing and sense-making in organizations. By and large, organizations may be conceived of as patterns of verbal interactions across individuals, communities, departments and divisions and between organizational tiers. Talking is what complements the written documents and protocol, and makes everyday practices work more smoothly, since it offers more detailed and contextual explanations and descriptions than written documents. The learning organization is always already a talking organization.

The concept of 'oral culture' is helpful when studying the construction industry because a substantial part of the communication is in fact verbal rather than written. While formal documentation is produced throughout the construction process, the communication in day-to-day work in architect offices, in design departments and on construction sites is primarily in the verbal form. As a consequence, theories about verbal communication are useful when examining operative work in the construction industry.

Project-based work

As is commonplace to point out in the construction management literature, the construction industry is a prime example of a project-based industry. Since there are only, by definition, non-routine production processes in the construction industry, there is a great need for effective management of the complex inter-professional and inter-organizational contractual and working relationships mobilized in a construction project (Bresnen, Goussevskaia and Swan, 2004: 1538). As a consequence, decentralized project-based organizations predominate in the construction industry, emphasizing the demand for handling the horizontal and vertical differentiation within construction firms. Speaking in terms of what Karl Weick (1976) calls *loosely-coupled systems*, both the industry and the focal construction firm tend to be loosely coupled (Dubois and Gadde, 2002). Boland, Lyytinen and Yoo (2007: 633) speak of 'complex building projects' as being representative of a network form of organizing. Construction projects are: (1) *distributed*, that is 'designed and constructed by multiple, autonomous actors'; (2) *heterogeneous*, composed of communities with distinct skills, expertise and interests; and (3) *sociotechnical*, 'trust, values and norms, as well as IT capabilities and complex fabrication processes'.

Speaking in more formal terms, the project organization form is designed to solve novel problems in creative ways and to perform task-specific activities that the traditional bureaucratic form is incapable of handling or is handling ineffectively. Expressed differently, projects are not of necessity adhering to standard operating procedures or the traditions of the line organization but are instead tightly coupled with their objectives and ends (Räisänen and Linde, 2004: 102). Working in a project setting demands mastery of both formal procedures and orchestrating activities that 'make things happen' and that 'work in practice' (Faulkner, 2007: 335; see also Fujimura, 1996, on the formulation of 'doable problems' in scientific work). Operating as a project manager in the construction industry demands significant formal training in the fields of accounting, line management, team building, and the ability to build and maintain networks of contacts, but also substantial experience from practical work building up a stock of know-how. Such know-how may include an 'appreciation of different types of client/user requirements, knowledge about specific products' relevant factors (for example, regulations), and useful contracts (suppliers, contractors, clients and other engineers)' (Faulkner, 2007: 335).

While there are significant benefits from using the project management model to organize and structure complex construction work, some researchers have suggested that there is a downside to the project organization. Perhaps the most important negative implication is that the project organization is sub-optimizing the firm's innovativeness, since individual project leaders are dedicated to meet their objectives and goals within restricted project budgets and therefore do not invest resources in innovation work. 'Because every project is unique and there are few possibilities for repetition', Drejer and Vinding (2006: 922) argue 'there is little reason for a building contractor to invest in innovation beyond the optimization of his own processes, which means that economies of scale and learning effects are largely absent'.[1] In general, practitioners and researchers alike debate the conservative attitude of the construction industry and its poorly developed innovative capabilities. These poor innovative capabilities derive from many practical, historical, cultural and legal factors, and it may be that the strong reliance on temporal organization is further undermining construction firms' capacity to innovate. Another problem with the project management model is that it is in fact less reliable and effective than its proponents tend to suggest. In the construction industry, there are a number of examples of grandiose fiascos, including the finalization of the Denver International Airport in Colorado, USA, including no less than four embarrassing postponements of the opening, a series of scandals and a final cost of $5 billion against the budgeted $1.5 billion (Cicmil and Hodgson, 2006: 7). Although it not, of course, the project organization form *per se* that is to be held accountable for such failures, there is nothing 'inherently effective' in the project-management organization as some proponents at times seem to suggest. Instead, the project-organization form has recently tended to be implemented not in domains characterized by complex and experimental work, but by reasonably standardized and routine-based work. In Räisänen and Linde's (2004: 117) account, 'In multi-project organizations today, projects are no longer the exceptional, unique and innovative work form of a new work order. Instead, project management is being subjected to the forces of organization rationalization, resulting in a bureaucratization of projectified activities'. Project management is instead becoming a standard organization form in post-bureaucratic organization fields.

In summary, the project-organization form predominates in the industry, providing opportunities but also imposing limitations for the actors in industry and derivative industries. Therefore, whenever studying construction work, the project organization form and its particular features and effects must be taken into account. Rather than being explicitly in favour or critical of the project-organization form, in this book, the project-organization form is perceived as one possible choice among many organization forms. As a consequence, the project organization form does not display any inherent qualities but can be developed into many different forms and variations.

Knowledge work in oral cultures: symbolism, materiality and tinkering

The construction industry is dominated by a series of characteristics such as its predominant reliance on verbal communication, its tinkering on the basis of material resources, and its strong occupational and professional cultures. The occupational groups include a variety of groups such as carpenters, electricians, plumbers, bricklayers and so forth, all of which have their own traditions – in many cases stretching back to the mediaeval guild system – specific skills, vocabularies, and other symbolic and cultural systems. The professional groups include engineers, designers and architects embedded in corresponding systems of material practices and symbolism. Construction work is, given all these heterogeneities and specificities, constituted by the ability to integrate these different domains of expertise into a coherent whole, an industry capable of producing durable and qualified goods and services. While most activities in the construction industry are at the bottom line aiming at producing some material effects, it is always a social production of buildings, landscapes, infrastructures (e.g. tunnels, roads, sewage systems), and so forth, deeply ingrained with socially embedded beliefs, norms and customs. When examining the use of knowledge in the construction industry, these social and emotional components need to be brought into the analysis. The various and local knowledge-management practices in the construction industry are already rooted in practices and beliefs, that is, they are what we in the final chapter of this book will refer to as institutions and the institutionalization of knowledge. Expressed differently, to understand how and why knowledge is used in construction work means to understand the broader complex of application and all its idiosyncrasies and situated and contingent knowledge.

Summary and conclusions

This first chapter of the book has aimed at outlining some of the challenges facing the construction industry and to point at the value of a knowledge-management perspective in dealing with these matters. A series of analytical concepts such as professions, institutions, and practices has been discussed. Thereafter, the literature on knowledge management has been reviewed in some detail, and some of the cultural and organizational features of the construction industry have been examined. In the next four chapters, empirical studies of how knowledge can be managed will be reported. Rather than adhering to an integrated and comprehensive theoretical model, these four studies acknowledge the heterogeneity and local differences between the four cases. In addition, the four chapters do not follow a shared structure or story-line but are structured to make the best sense out of four individual cases.

2 Site management work and the value of coaching

A relational view of knowledge

Introduction

The aim of this chapter is to explore how middle managers in the construction industry conceive of their work life and to identify some of the major challenges that site managers are grappling with. This is of interest from a knowledge-management perspective because site managers play a central role in managing, directing, monitoring and leading construction projects. Site managers strongly influence the effectiveness of the construction project, therefore, the support and help directed towards site managers are part of a broader knowledge-management system helping construction projects exploit its accumulated knowledge. However, the point of departure for this study is the rather negative view of middle management in much management literature. Research on middle managers remains one of the most marginalized fields of interest in management studies (Floyd and Woolridge, 1997; Thomas and Linstead, 2002). In comparison to the massive literature on leadership, the body of texts on middle managers is miniscule. Moreover, in comparison with the so-called labour process studies tradition, emphasizing the worker's position and everyday work-life experiences, the middle manager is a marginal figure. In cases where middle management is examined, it is portrayed in the bleakest of terms. Middle managers are on the way out, continually being reduced in numbers when organizations are rationalized and adopting a flatter structure. They are stressed and burned out, and endure a work-life situation wherein they are stuck between the demands of top management and those of their co-workers and subordinates (Dopson and Stewart, 1990).

In the construction industry, it is site managers at construction sites who play the important middle-management role. For instance, Mustapha and Naoum (1998: 1) write: 'The site manager stands at the heart of the success or the failure of the project for the contractor, the professional team, the client and ultimately the general public'. Site managers have a work assignment that is notorious for being based on their ability to combine a variety of heterogeneous activities, both practical and administrative, and for demanding a significant amount of work from the site manager in question. Studies of site managers suggest that they experience their work life as problematic in terms of needing to juggle a multiplicity of activities and because they have to endure a work situation wherein it is essentially problematic to predict forthcoming events and occurrences (Davidson and Sutherland, 1992; Mustapha and Naoum, 1998). Thus, site managers' work often requires them to navigate situations fraught with ambiguities and emerging events and challenges.

This chapter presents a study of how Swedish site managers regard their position between, on the one hand, the demands of top management and the expectations for financial performance and smooth-running project management activities and, on the other hand, their co-workers in the projects. The study suggests that rather than seeing middle managers and site managers as an overtly disfavoured group, unable to establish a work situation wherein they can mediate expectations from different stakeholders and colleagues, one must recognize the level of professional skill and work experience required to operate as a site manager. In other words, site managers' work needs to be examined in more positive terms than is generally suggested in both the literature relating to site managers and in studies addressing the role of middle management.

In the latter half of the chapter, a study aimed at exploring the value of coaching for site managers is reported. Since site managers are central actors in the day-to-day work at a construction site, the coaching of this particular group is arguably a form of knowledge management that, contrary to more formal education and training, resides in day-to-day work experiences. As a consequence, coaching is potentially helpful for site managers and other construction industry representatives handling a large set of decisions, assignments and objectives on an everyday basis.

Setting the scene: site managers and their work

The actors: middle managers and site managers

In the following there are two different strands of research that need to be attended to: first, literature addressing middle managers in large hierarchical organizations; and second, publications discussing the role of site managers in construction projects. The 'general middle-management literature' tends to focus on more conventional organizations and firms rather than those construction companies that of necessity have developed their own rather idiosyncratic form of middle manager – the site manager. Conversely, the literature on site managers does not make reference to management literature reporting studies of companies outside the construction industry. This separation between the construction industry and other industries is problematic because it portrays the construction industry as different from other organizational activities. When bridging the two traditions of research, one may present a more integrated and complex picture of how middle management work is constituted in organizations.

Dopson and Steward (1990) point at the rather negative image of the middle manager's position in the firm:

> Few people have anything encouraging so say about middle management … most people portray the middle managers as a frustrated, disillusioned individual caught in the middle of a hierarchy impotent and with no real hope of career progression. The work is dreary, the careers are frustrating and information technology, some writers argue, will make the role yet more routine, uninteresting and unimportant.
>
> (Dopson and Stewart, 1990: 3)

For some reason, such critical accounts of middle management prevail in the literature. For instance, Thomas and Linstead (2002: 3) say that in the existing literature on middle management 'managers are portrayed as univocal and homogeneous entities that are passive victims rather than active agents constructing, resisting and challenging the subjectivities offered them'. For Thomas and Linstead (2002), middle managers are often envisaged as agents that have neither the capacity nor the mandate actively to influence the tactics and strategic decision-making of the firm. Instead, they are assigned an intermediary role in between the 'thinking' (top management) and the 'doing' (subordinate workers). However, there are some examples of how middle management is portrayed in more positive terms. King, Fowler and Zeithaml (2001) argue that middle managers play a central role in implementing strategic decisions on the shop-floor and that middle management very much remain an untapped reserve in firms:

> Middle managers play an essential, but often unappreciated, role in successful strategy making. Middle managers' participation in strategy formulation is associated with improved firm performances, and their commitment is critical to successful strategy implementation.
>
> (King, Fowler and Zeithaml, 2001: 98)

In a similar vein, Delmestri and Walgenbach (2005) argue on the basis of research on middle management work in the UK, Italy and Germany that middle managers play either the role of knowledge brokers (as in the case of the UK) or as technical experts capable of practical problem-solving. In the three cases, middle managers thought of themselves as gatekeepers serving top management, responsible for dealing with small and everyday concerns, thereby preventing them from moving up the hierarchy and stealing valuable time from top management. Other writers such as Nonaka and Takeushi (1995), drawing on a Japanese management model emphasizing middle management as central to a firm's performance, suggest a 'middle up–down' strategy, wherein middle managers, who have a detailed understanding of operations yet are capable of seeing the broader picture, should have a say in the formulation of strategy. Similarly, Huy (2002) argues that middle managers are capable of dealing with emotional responses among the co-workers during organization changes, and therefore play a central role in organization change programmes.

In summary, middle managers are relatively marginalized in terms of research interests and in many cases are portrayed in unnecessarily critical terms. However, they have also been considered as a central group for indicating a firm's performance and implementing strategic decisions.

Site managers' working life

The literature on site managers in the construction industry is almost as negatively slanted as the middle management literature. Again, site managers are portrayed as a professional group exposed to conflicting demands and objectives and a work situation complicated to handle. For instance, Djerbarni (1996: 281) writes:

> Site managers carry out one of the toughest and hardest jobs in the construction process. Site management is characterized by a high work overload, long working hours, and many conflicting parties to deal with including the management, the subcontractors, the subordinates, the client, etc. This trait of the job makes it very prone to stress.
> (Djerbarni, 1996: 281)

Djerbarni (1996) examines how stress influences the work of site managers and suggests that the presence of stress and burnout in site manager work remains one of the main challenges for the industry. In another study, Davidson and Sutherland (1992: 33) claim that site managers in the construction industry are more exposed to stressors than managers in other industries: 'Levels of reported job satisfaction were significantly lower for managers in the construction industry than among managers and supervisory grades employed in engineering'. Among the most important stressors were 'time pressure' and 'working long hours'. Furthermore, Davidson and Sutherland (1992) point at the amount of 'paperwork' as one of the most important sources of stress. Mustapha and Naoum (1998) examined the factors that determine site manager effectiveness and found that personal qualities and job conditions such as job satisfaction are the two most important. Effective site managers, the findings suggest, are then primarily high performers because of their individual skills.

Fraser's (2000) study of site managers' work supports this emphasis on personal characteristics, and suggests that they strongly affect the performance of the construction project. Fraser thus sides with a number of leadership studies emphasizing the charismatic element in all leadership work (Steyrer, 1998; Ball and Carter, 2002; Flynn and Staw, 2003).

However, the image of the work situation of site managers sketched by Djerbarni (1996), Davidson and Sutherland (1992) and Fraser (2000) largely supports the findings from studies of leadership work in other settings. Ethnographic accounts of leadership presented in Carlsson's (1951) seminal study and later in Mintzberg's (1973) and Tengblad's (2002) work testify to a fragmented working day, based on numerous short interactions with co-workers and continuously interrupted by minor issues that needed attention (Kotter, 1982).

However, it is worth taking into account some of the specific characteristics of the site manager's work. First, the site manager is responsible for not only the technical and production-oriented matters at the construction site, but must also be trained in administrative work, juridical aspects, human resource management and other areas of expertise that are generally functionally organized into different departments and work roles in firms. To a greater extent, the site manager needs to be more of a 'jack-of-all-trades' than middle managers in other industries. Several studies of construction work suggest that experiential learning is one of the most central learning strategies in the industry (Ogulana, 1991; Lowe and Skitmore, 1994; Love, Li, Irani and Faniran, 2000). This also applies for site managers, whose work demands significant experience. Second, site managers are often working on their own with few if any colleagues (i.e. other site managers) in their immediate proximity. Therefore, site managers have fewer opportunities for collaborating with peers. Because of the demands for mastering a multiplicity of processes and activities and the specific work situation with few colleagues, site managers are likely to find themselves stuck between not only the expectations of top management and construction workers, but also conflicting priorities and objectives and goals. Third, some researchers suggest that conflicts and controversies are endemic in the industry, thereby locating the site managers in a situation requiring constant renegotiation of contracts and relations, and resolution of disputes. For instance, Rooke, Seymour and Fellows (2004: 655) argue that there is evidence that the UK construction industry is 'opportunistic, prone to conflict and resistant to change and that these characteristics impede competitiveness and overall efficiency'. Latham (1995) referred to 'a culture of claims' and noted 'the industry has deeply ingrained adversarial attitudes ... the culture of conflict seems to be deeply embedded' (Latham, 1995: 5). Maybe partly due to cultural traits but also derived from the project organization form dominating in the construction industry, site managers have to be able to weigh interests and priorities against each other. Their everyday work is rarely, if ever, devoid of unanticipated events or conflicts. However, in order to understand how site managers conceive of their role and their work-life opportunities, systematic studies of how site managers cope with this highly demanding work situation are required. Rather than adhering to positivist and qualitative methods (see e.g. Davidson and Sutherland, 1992; Djerbarni, 1996; Mustapha and Naoum, 1998), qualitative methodologies (e.g. interviews) may shed further light on the site managers' work-life experiences.

Project work, gender and the construction industry

The white masculine domination in the construction industry

The construction industry is, when including all contractors and derivative industry suppliers, the second largest industry in Sweden after the health-care sector. From the outset, the construction industry has been characterized by being associated with masculine virtues and skills (Appelbaum, 1981) and for being organized in the form of projects (Green, 2006). These two traits will be examined further in the following text.

Historically, the need for physical strength and the lack of adequate tools and machinery have made construction work an occupation largely dominated by men. To date some women are educated and trained to work in the industry, but they are often appointed to managerial positions or play the role of technical experts in architecture and design work. Not many female construction workers, foremen – itself a gendered term – and site managers work in the industry. 'Women are significantly under-represented in construction craft training programmes', Agapiou (2002: 698) notes. In her study of female civil engineers, Faulkner (2007: 334) found that 'virtually all women interviewed *had a story to tell* about why they made the choice – in much the same way as women who don't have children have a story to tell as to why: it demands an explanation'. Being a woman in the construction industry is, Faulkner's (2007) female interlocutors thought, something that demands justification and a proper explanation. Based on a study of the view of female construction workers among both male and female Scottish construction industry representatives, Agapiou (2002) reports some remarkable statements on part of the male skilled workers, including 'women don't have the innate ability to use the tools', 'they don't have the natural understanding of building that men do' and 'women aren't designed to lift heavy material'. Fowler and Wilson (2004) found similar sexist beliefs among male architects in a study in the UK:

> Some [male architects] offered psychological explanations for women's subordinate position in the profession, such as the view that women were weaker in 3D perception ... Other male architects identified fitting jobs for women in less spectacular areas of design – in domestic architecture or interiors.
>
> (Fowler and Wilson, 2004: 114)

However, not all interlocutors in Agapious's (2002) study expressed such outright sexist beliefs but instead expressed a more gentlemanly yet equally problematic attitude, claiming that women would be given the easiest and least heavy tasks. For this group, trades such as painting, plastering, tiling, joinery and electric fitting were those 'most likely to be mentioned as "particularly appropriate" for women', Agapiou (2002: 702) reports. Some interlocutors unreservedly welcomed more women to the industry, expressing no particular gendered beliefs. Taken together, the resistance to women in construction is 'based largely on folklore, fears and fallacy' (Agapiou, 2002: 704). While the social costs derived from the patriarchal structure of the industry are under-researched to date, the survey research of Sang, Dainty and Ison (2007) among female architects in the UK offers some indication of the state of the industry:

> The data presented here suggest that women working within the architectural profession are at greater risk of poor health and well-being as a result of occupational stress. Overall female architects appear to experience lower job satisfaction, poorer physical health, higher work-life conflict and higher turnover intentions.
>
> (Sang, Dainty and Ison, 2007: 1314)

On the basis of these findings, Sang, Dainty and Ison (2007) speculate whether gender may be regarded a 'risk factor' in the industry. Fowler and Wilson formulate an equally daunting conclusion:

> [T]here are few grounds for the belief that women are on the verge of 'making it' in architecture, It is not that women lack the cultural capital to do well in the profession, for nobody has doubts about their ability at architectural school level. Rather, we suggest that where markets are less localized and less forthcoming, the room for tolerance and nurture of those with young children becomes reduced. More specifically, in a savagely competitive climate, contracts place a strong premium on instrumental rationality, not least in the use of power to insist on the time discipline of builders and others.
>
> (Fowler and Wilson, 2004: 116)

There is little empirical evidence suggesting that the Scottish case reported by Agapiou (2002), the studies of Sang, Dainty and Ison (2007) and Fowler and Wilson (2004), and those in architects' work, would not be representative of gendered views of construction in other countries and regions. In addition, the degree of diversity in the construction industry is low. In April 2005, the Swedish Integration Ombudsman reported that the number of workers in Swedish manufacturing industry born outside of Sweden was in fact lower today than at the end of the 1980s. Representatives of the industry as well as researchers tend to portray the industry as conservative and unwilling to adapt to new conditions (Kadefors, 1995; Huemer and Östegren, 2000; Dubois and Gadde, 2002). The industry is in general strongly affected by masculine values and norms (Gherardi and Nicolini, 2002) and construction companies are often modestly interested in adopting new work methods and new techniques. Dainty, Bagilhole, Ansari and Jackson (2004) point out that the representation of women in the industry is a meagre 10 per cent of the workforce even though women constituted 46 per cent of the British workforce in 2001.

Regarding ethnic minorities, less than 2 per cent of workers in the construction industry could be categorized as having an alternative ethnic background. In the UK, about 6.7 per cent of the workforce belongs to this category. Not only were ethnic groups and especially women under-represented in the industry, Dainty, Bagilhole, Ansari and Jackson (2004: 79) report that both women and black and Asian construction managers had experienced discrimination and harassment (in the case of female managers) and 'racist name-calling, jokes, harassment, bullying, intimidation, and physical violence' (in the case of black and Asian managers). To overcome these obstacles for a more diverse construction industry, the authors also suggest a series of changes for the industry including closer monitoring of individual manager's careers and a culture change in the industry. Loosemore and Chau (2002) provide equally negative results from the Australian construction industry and suggest that construction companies essentially fail to exploit the 'positive attributes of multiculturalism':

> Overall, our results suggest that racism is seen as an inevitable consequence of working in the construction industry and one that is largely ignored by managers and accepted and tolerated by workers.
> (Loosemore and Chau, 2002: 97)

However, not all studies discuss the industry in such negative terms. English (2002) reports a study of a training programme in a major South African construction company and shows that training can enhance the efficiency of the construction company and therefore strengthen its competitive advantage. In the only study providing evidence from the Swedish construction industry, Dadfar and Gustafsson (1992) point at the importance of effective management practices in international construction projects. They argue on the basis of research on construction projects in the Middle East that, in many cases, Scandinavian construction managers have limited or little competence in managing diverse and cross-cultural projects. Women and especially ethnic minorities are poorly represented and, in cases where there are construction workers of a different ethnical background, controversies may appear.

The gendering of project work

The other significant characteristic of the construction industry is its historical and actual reliance on *temporal organization*, that is, project work organization (Lundin and Steinthórsson, 2003). Comparisons between manufacturing industry and construction work suggest that, while the former industry has to some extent de-skilled the workers, the latter essentially remains a 'craft-based industry' (Stinchcombe, 1959), leaving the co-workers as skilled craftsmen maintaining the control over their work. In addition, proponents of project management organization (e.g. Keeling, 2000) regularly use construction industry and major civil engineering projects as examples of the effective deployment of the project form. However, even the early project-management literature (Wilemon and Cicero, 1970; Butler, 1973) points out that the project form and its integration of a series of complex activities leads to conflicts of interest that the project manager needs to resolve. While the normative or mainstream project-management literature tends to perceive project work as being the ideal organization form when managing complex undertakings, a more recent and more critical project-management literature emphasizes the bureaucratic and control-oriented features of project-management work. For instance, Hodgson (2002, 2004, 2005) and Räsäinen and Linde (2004) have pointed at the return to and re-articulation of Taylorist and rationalist management principles in the normative project-management literature. While project management has gradually established itself as a sub-discipline of the management literature, it is today, when facing its maturity, examined in terms of having a series of concerns to address (Hobday, 2000; Söderlund, 2004). One such challenge is how to promote learning within and especially between projects (Bresnen, Goussevskaia and Swan, 2004). More specifically, the very foundation of the project management form and practice, the institutionalization of legitimate and jointly agreed upon principles, has been subject to analysis. Hodgson and Cicmil (2007: 445) advocate such an analysis:

> As project management relies upon the naturalization of 'the project' itself as both focus and *raison d'être*, to critique project management we must start to question the ontological foundations of 'the project', drawing on perspectives which would instead see 'the project' as a constructed entity, with powerful and often unrecognized consequences for the management of what we label as 'projects' in contemporary organizations.
>
> (Hodgson and Cicmil, 2007: 437)

By examining the American association Project Management Institute's (PMI) project *Book of Knowledge (PMBOK)*, Hodgson and Cicmil (2007) argue that the recommendation and 'standards' advanced by PMI as being uncontroversial and mutually beneficial for a variety interest groups are in fact relying on a rather specific rationalist and instrumental world view, portraying project management work as a series of activities aimed at controlling and monitoring the work process in detail. PMI is attempting to play the role of the global institution endowed with the legitimacy and competence to monitor and control project management practice, and therefore its directives need to be subject to critical analysis and discussions. Buckle and Thomas (2003: 433), declaring that project management 'has been characterized as a "macho profession"', examine the gendered features of the *PMBOK*. They claim that documents such as *PMBOK* need to be examined because they 'strongly influence the development of the emerging profession' and in terms of being 'a potential force for isomorphism, signaling project managers worldwide about the appropriate use of masculinity and femininity' (Buckle and Thomas, 2003: 439). In their analysis, they conclude:

> By exploring the symbolic language indicators of appropriate project management behaviour we suggest that the hard masculine systems exert considerable influence on the 'best practice' outlined in the *PMBOK*. Softer feminine logic systems appear less influential and presumably less valued or trusted in the profession.
>
> (Buckle and Thomas, 2003: 439)

Studies of actual projects in industry report similar findings regarding the gendering of project work. Lindgren and Packendorff (2006) studied an information technology project in a Swedish company and concluded that project-management practice is not a value-neutral or disinterested work procedure but demands the full commitment from project co-workers:

> In a time when the traditional masculinities of managerial work are subject to a lot of debate and criticism in society ... project work seems to be a way of reintroducing many of them in the guise of short-term efficiency. An individual constantly involved in demanding projects work will be as separated from her/his family life and emotions as managers and entrepreneurs have always been.
>
> (Lindgren and Packendorff, 2006: 862)

In addition to the critique of project form and its alleged 'masculine' conception of social reality, ethnographic studies of construction firms suggest that the industry is dominated by masculine values and norms.

In summary, project-management work is often presented by normative and mainstream writers as being a handy solution to a range of managerial and organizational concerns in complex undertakings. Contrary to such a view, more critical accounts of project organization suggest that project-management practice represents a return to a Taylorist view of work and organization. In Hodgson's (2004: 86) straightforward formulation, 'what distinguishes project management as of particular relevance to 21st-century organizations is its rediscovery of a very 19th-century preoccupation with comprehensive planning, linked to a belief in the necessity of tight managerial control'. In addition, neither project work nor the (normative) project-management literature is gender-neutral but in fact encourages practices, values, norms and beliefs that are stereotypical masculine and only limited attention is paid to supposedly 'feminine' qualities.

Site managers' everyday work in construction projects

In the following, a number of site managers account for their work as being the muddling through of a series of heterogeneous activities. Being able to navigate in such complex domains was regarded as the primary competence of the site manager, and such skills and experiences were highly prized by the site managers. In addition, they thought of these skills as being of central importance for the mastery of their work and performance of the construction project.

Day-to-day work practices

Among the site managers interviewed, all of them appreciated their job and the freedom the work assignment entailed. The site managers pointed at the creative part of the work – that they actually contributed to the production of a building. One of the site managers (Site manager 4) said: 'What's most positive is that from this very blueprint and a time line, you see something grow out there, a physical effect. I consider that very rewarding'. The site managers also pointed at the need for being at the centre of the operations. One site manager argued: 'I, in the role of being site manager need to know what is going on at the site ... I coordinate and thus I need to know what everybody is doing' (Site manager 3). Another site manager said: 'The site manager is the hub around which everything revolves' (Site manager 4). Therefore, there was a great demand on the site manager to forge good relationships with the co-workers. Site manager 3 continued: 'I always spend time eating with the lads in the cabin. I try to maintain a good contact and good communication. I don't lock myself up in my room to keep them away'. The other site managers tended to agree on the importance of communication skills:

Q: How do you create confidence in the workplace?
A: I think you need to demonstrate to the lads that they are needed.
Q: And how do you do that?
A: It depends how you talk to them. You speak in a certain manner. 'Now, you do like this', a bit like 'hierarchical ordering' if you like, then you do not get the response you asked for. Instead, you go 'I need help to fix this and that and I believe you can help me, right?' and then he thinks that he's really making a difference. In the same manner, if they are ill they expect you to call them to ask how they are doing. That is very

important, especially regarding the lads who are ill quite often. You need to know whether it is somebody in the workplace being responsible [for the illness].

(Site manager 7)

Another site manager argued:

I think you can go a long way if you are skilled in talking to people. You mustn't stay on your own and believe you know everything. There's always somebody who knows better than you. As a site manager you need, if not a safety net, then a number of people around you that you know that you can call when things go wrong. That is very important.

(Site manager 4)

He continued:

A great deal of common sense is what you need; that you greet everyone, that you walk around and ask how people are and dedicate the time needed for it, then I think you've gained a lot. If you're stressed out and they call and tell you that 'now, it all went down the drain', and then you run around and pass fifteen workers and you don't even bother to say hello – very negative! That's really not good at all.

(Site manager 4)

In other words, the site manager needs to serve as the leader of the site, not only as an administrator. This double role adds significantly to the workload of site managers. For instance, the amount of meetings was at times regarded as a problem:

Q: Are there many meetings?
A: Well, it's outrageous! But it may be that it is extreme out here. They're [the client organization co-workers] crazy about meetings: you even have to prioritize meetings, or otherwise you spend the whole day at meetings.

(Site manager 1)

In everyday work life, site managers tended to think of their job as being stressful and demanding. One of the site managers argued that stress was contingent and cyclical:

Q: Is it stressful?
A: Yes, you may say that. Now it's very cyclical; some periods you may say that what you think of it as being at a normal level, that is it acceptable, but then there are periods when you really ask yourself what you're doing.

(Site manager 1)

One of the foremen, a former site manager, agreed:

> It is stressful. And you have this feeling you never get a fair chance to finish anything. You may have like ten different jobs going on at the same time and you never complete them. In many cases you have to use the night to shovel off things so you can start off fresh with the next thing [in the morning].
>
> (Foreman)

One site manager emphasized the tendency in the industry to favour 'lean production' forms of organizing new projects:

> The construction projects get shorter and shorter, and that's a negative thing ... the real estate companies save some money that way ... you don't let a project team of four persons run such a project a year in advance. Instead, you start the design work more or less at the same time as we start the production. It is a really slim organization these days, especially in this type of contracting [total contracting].
>
> (Site manager 5)

One of the foremen added:

> I believe it is not that easy to please everyone. He's [the site manager] under pressure from the company to make everything run smoothly and we have to make some money on every project ... Today, the time lines are so compressed, so I believe he really needs to take a great deal of responsibility to find solutions in order to keep the schedule.
>
> (Foreman)

One site manager pointed at the work pressure as being on the verge of what was possible to handle:

> Today, with the degree of support I have here, there is no chance that I can take on any more work. I do not even handle what I have as it is. We have pointed out that there is a need for more support.
>
> (Site manager 3)

He continued:

> You need to be able to cope with stress. You need to remain calm. The most important things is that, even if it may be stressful, you need to think that 'it cannot be that important' and then you need to take the time to think things through rather than doing some things offhand just because you think it's panic.
>
> (Site manager 3)

ADMINISTRATION VERSUS PRODUCTION RESPONSIBILITIES

The primary source of the stress was the trade-off between different priorities, in most cases production issues versus administration issues. One of the site managers argued:

> You are always split between production and administration. I think after all that I enjoy a bit of both but what's demeaning is that you are always split in half. You often do not have the time to engage carefully with either of the two, but you make two things equally mediocre.
>
> (Site manager 1)

As a consequence, there was a continual need for making priorities:

> What I prioritize most of all is that the production runs. If that works, we keep to the time schedule and everybody gets the chance to do their work. The administration is less prioritized. I'd rather take some complaints because of that than delays in production. Administration can always be dealt with afterwards. You need to deal with what happens here and now. If that does not run as intended, you never catch up.
>
> (Site manager 1)

This view was shared among all the site managers:

> The production [is prioritized]: to get a smooth and comfortable process at the site so you can make things in the right order and can avoid panic situations. It's supposed to run smooth and that is a matter of planning.
>
> (Site manager 2)

> If I do not have the time, then I prioritize the production. If that does not work, the economy goes with it.
>
> (Site manager 3)

One of the site managers thought of this new role as being of rather recent origin:

> The old site manager role, back then, was more emphasizing the production management. I know that back then, you did not even see any invoices out there. Somebody else took care of them. What has happened during the years is that more and more administrative work has been decentralized and transferred to the projects while just as much production management work is still expected. This makes your work fragmented.
>
> (Site manager 1)

This recent site manager role implied, *inter alia*, that a significant amount of paperwork was brought out to the construction sites from the central office. This situation is problematic for two reasons in the eyes of the site managers. First, many competent potential site managers are not overtly attracted to the site manager job because of its administrative outline. One argued: 'A lot of people really do not want to deal with paperwork; they want to spend time on the site. They may have the competence to build houses but not to take care of paper work' (Site manager 4). Additionally, the office workspaces provided on the construction sites did not promote good conditions for dealing with paperwork. One site manager explained:

> In a place like this, it is really hard to close the door and say that 'I need to be on my own for a while, because I have a protocol or I have some things to take care of so I need a bit of calm'. Then the telephone rings, people come to see you all the time. But after four o'clock, then things calm down and then you can gather all your thoughts.
>
> (Site manager 4)

Another site manager emphasized the need for a more modern view of the site managers' needs regarding office space:

> You live the life of a gipsy ... These barracks are rented, and we change them all the time. I suggest that we should have a bit more personal offices that we take with us from site to site. Then you are able to have an office; you know where your things are and you can keep it neat and tidy in a completely different manner.
>
> (Site manager 3)

The combination of a demanding administrative workload and the lack of adequate work conditions was specifically cumbersome in smaller construction projects where there was little opportunity for private offices. For some of the site managers this was a major concern, while for others it was less important. Another concern for the site managers was that in some cases their line managers did not offer adequate support. In most cases, this worked without major conflict, but there were some clashes of interest between site managers' and line managers' objectives:

> If I have the ambition to take care of things in a certain manner and be able to deal with my job assignments, then I get the response that 'you may need to lower your ambition. We cannot afford to work at that level'.
>
> (Site manager 3)

In other cases, the line managers were regarded as concerning themselves too much with details in the site manager's territory: 'He [the line manager] does not really bother with the details ... Some of them have been too much into details and then you think that they should stay out of it' (Site manager 1).

Concluding remarks

The site manager's job was broadly appreciated by the interviewees who liked the 'creativity' of the work and the unpredictable nature of construction work. At the same time, they felt there were conflicting interests in terms of imposing stressful work situations and the continual need for making decisions on what activities to prioritize at each moment. Even though the site managers' work was regarded as based on certain standard operation procedures, the interviewees emphasized the contingent nature of construction projects.

Q: What are desirable qualities of a site manager?
A: That's a tricky question. I think there can be rather different qualities depending on the project. If you work on you own in a small project, then you need to be really skilful in production matters, preferably know all of it. If you are the site manager of a 100 million [in Swedish crowns, approximately 10 million euros] project, then it may be a disaster ... there you need to motivate people and delegate tasks to the foremen available ... We've seen several cases where brilliant site managers are assigned a major project and then they stick to their routine and then it all ends in a crash because you cannot monitor everything in detail in this case.

(Site manager 6)

One of the site managers concluded: 'One site manager differs from another. We're very different. I'm sure we work in a variety of ways. You mustn't believe that everyone is as you expect them to be' (Site manager 4). The site manager's work is portrayed here as series of hard-learned skills embedded in personal experiences. 'Diverse work demands diverse competencies', the site manager suggested.

Managing knowledge through coaching: an explorative study

The concept of coaching

In this part of the chapter, a study of the use of coaching to support site managers in their day-to-day work is reported. The methodology of the study is provided in the appendix of the book. In this section, the literature on coaching is considered; thereafter, empirical material is examined.

Although coaching, and the more narrow term 'executive coaching', is one of the recent buzzwords in the media and popular press denoting a variety of procedures wherein individuals receive advice from and discuss their concerns with some authority in a specific field, there is a surprisingly limited management literature on coaching (Gray, 2006). Most of the available reports explain what coaching is or detail successful coaching projects (Levinson, 1996; Peterson, 1996; Hall, Otazo and Hollenbeck, 1999; O'Shaughnessy, 2001; Palmer, 2003; Wasylyshyn, 2003). There is almost no critical or analytical literature addressing the underlying assumptions and beliefs to any coaching procedures (see Berglas, 2002, for an exception). In addition, the bulk of the literature is written by individuals with vested interests, that is, professional coaches presenting their work procedures or methods, or reporting successful show cases (Burdett, 1998; Arnaud, 2003; Hackman and Wageman, 2005). In a study of Australian coaches and their efforts to construct viable and legitimate identities and roles in the market, Clegg, Rhodes and Kornberger (2007) found that coaches were eager to define themselves in negative terms, that is, in terms of what they were *not* – for instance, management consultants. They conclude their study:

> As an industry, business coaching is ill-defined, contradictory and ambiguous. Indeed, it is this apparent lack of an established order within which coaches work that enables them to try to construct their organizational identities. By this account, organizational identity is not an essence of a substance fleshed out by characteristics; rather, organizational identity is enacted and embedded in a field of differences.
>
> (Clegg, Rhodes and Kornberger, 2007: 511)

Coaches are apparently not fully established to date as legitimate actors in the organizational field. There is a range of definitions of coaching in the literature. Kilburg (2000, cited by Kampa-Kokesch and Anderson, 2001) defines coaching accordingly:

> A helping relationship formed between a client who has managerial authority and responsibility in an organization and a consultant who uses a variety of behavioural techniques and methods to help the client achieve a mutually identified set of goals to improve his or her professional performance satisfaction, and, consequently, to improve the effectiveness of the client's organization within a formally defined coaching agreement.
>
> (Kilburg, 2000: 67, cited by Kampa-Kokesch and
> Anderson, 2001: 208)

Kampa-Kokesch and Anderson (2001: 210) emphasize the scope of coaching, suggesting that it is more 'issue-focused' than therapy and that it occurs in a 'broad array of contexts' including 'face-to-face sessions, meetings with other people, observations sessions, and by e-mail'. For Hall, Otazo and Hollenbeck (1999: 40) coaching is meant to be 'a practical, goal-focused form of personal, one-on-one learning for busy executives'. The strength of coaching is, Hall, Otazo and Hollenbeck (1999) continue, that it 'provides executives important feedback that they would normally never get about personal, performance, career, and organizational issues'. Peterson (1996: 78) defines coaching in more general terms as 'the process of equipping people with the tools, knowledge, and opportunities they need to develop themselves and become more effective'. Many writers underline that coaching differs substantially in time and space and that there is a variety of methods and tools in use: 'Executive coaches vary significantly in terms of the methodologies, approaches, tools, and durations of their executive coaching engagements', Wasylyshyn (2003: 102) claims. More specifically, issues covered by executive coaching programmes may range from 'easy ones' such as 'writing skills, setting priorities and assessing staff needs' (Hall, Otazo and Hollenbeck, 1999: 40) to more difficult problems such as 'improving relationships with bosses, improving interpersonal skills, and how to implement layoffs'.

Another important distinction in the literature is that between the coach, the mentor, and the counsellor (Gray, 2006: 476-7). A mentor is here a person internal to the firm, in many cases in an executive position, supporting and helping his or her protégé to handle various practical and political concerns (Judge and Cowell, 1997: 71–2). A counsellor, on the other hand, is a specialist trained to handle more complicated psychological problems. Seen in this view, the role of a coach falls between that of the more informal assistance of the mentor and the more advanced and scientifically grounded support of the counsellor. However, even though coaching may promise a lot, critics contend that there is a risk, that the more complicated problems of those receiving coaching are being ignored, or dealt with inadequately by executive coaches. Berglas (2002: 87) argues:

> By dint of their [coaches] backgrounds and biases, they downplay or ignore deep-seated psychological problems they don't understand. Even more concerning, when an executive's problem stems from undetected or ignored psychological difficulties, coaching can actually make a bad situation worse.

Seen in this view, coaching is not a universal remedy for all kinds of organizational and managerial malaises, but as proponents of coaching point out, must be carefully designed, executed, monitored and evaluated to accomplish the managerial objectives. When carefully designed and managed, a coaching programme is a strategic resource in the firm's human resource management practices. Failing to run a coaching programme professionally may lead to disappointment, cynicism or alienation on the part of the coachees. Still, coaching is potentially useful for sorting out roles, processes and relations, and for making managers actively engage in continuous self-evaluation of and reflection over their priorities and leadership skills. It is this latter role that coaching is supposed to play in the present research project. Site managers in six construction sites were selected to participate in a research project conducted by Chalmers University of Technology.

Coaching site managers: learning, outcome and criticism

At the start of the coaching process, some of the six site managers expressed some concern regarding the additional workload the coaching activities would lead to and their own ability to take advantage of the coaching process. In addition, they were concerned about what they considered a lack of transparency of the entire process: they could neither see the objective of the coaching activities, nor were given a roadmap they could rely on. However, after the first few weeks, when they had got to know the coach and met with their fellow coachees, they gradually came to appreciate the opportunity of taking part in the coaching activities. At the end of the yearlong process, all the site managers could point out some specific skills and competencies they thought they had improved during the period. One of them argued that he had improved his skills for reflecting: 'what I really learned is this thing about reflection, to take the time to really think about what happened after a conflict or a meeting, or any other event' (Site manager 1).

Such reflection skills are important for practising site managers because their work entails continuous meetings and encounters with individuals, in some of which they will either deal with conflicts or diverging views or produce conflicts. In such situations, as a site manager said, 'it is quite common that our work becomes emotional. ... We are handling a range of meetings and there are a certain amount of conflicts. Thus it is a clear advantage to be able to recognize an emerging conflict ... and how to deal with conflicts' (Site manager 1). Conflicts could be with employees, sub-contractors or clients, and failing to reflect on how such situations emerge and develop and one's own role may lead to an unnecessary continuation of the conflict. One site manager told a story wherein he thought he could use the methods from the coaching process to handle a situation better. In a construction project, one of the sub-contractors wanted to re-negotiate the contract but the site manager thought the sub-contractor could only blame himself for failing to deliver the work at specified costs. However, as a new negotiation eventually took place, the sub-contractor managed to obtain approval for his new contract proposal. Thus the site managers thought that he 'had lost the case'. 'That really affected my ego in a negative manner, you may say', he admitted. Prior to the coaching experience, the site manager would have found it difficult to forget such a setback but now, having been trained and encouraged to conceive of a situation in new terms, he thought he could see the whole situation from various angles. Another site manager expressed a similar idea about being better equipped to take on the role of the other and to see different points of view on a matter:

> Before, I thought certain things were bleeding annoying; you think you have provided clear directives and yet they were not followed, and so forth. I have come to realize that maybe you need to change yourself and the way you are presenting the directives. Then you need to understand the objectives of your co-workers as well. Previously, I used to think that here we are, all of us, hired to do a job, all of us share the same ideas, but that is not necessarily the case.
>
> (Site manager 3)

Rather than criticizing the co-workers for failing to follow his instructions, the site manager put much effort into showing and telling in greater detail how and why he wanted the work to be carried out:

> Once I had this picture of the situation, I explained in greater detail what needed to be done and delegated various tasks to the foremen and were more clear about the objectives of the tasks. One may even say that I was overworking my instructions. But that gave an immediate effect, I think, and the end result was better.
>
> (Site manager 3)

Today, the site manager argued, his co-workers thought of his communication as being more detailed and more easily understood. He even took the time to explain what he thought was 'the basics'. He contended: 'I think this [detailed instructions] has been appreciated even though I was concerned that the opposite could have been the case' (Site manager 3).

One of the younger site managers expressed his newly improved skills derived from the coaching experience in terms of being 'a better listener'. He explained that he had become:

> A little better at listening ... [having] a little better understanding of my own role ... Even if I do exactly as I used to do, throwing out comments just like that, at times I may stop for a while and think 'Jeez, that wasn't a particularly smart thing to say', or 'that actually worked very well'; being able to take one step back and look at myself from the outside to realize that 'I could actually have been able to influence that situation rather than doing as I always do'. I have not developed so far that I am able to stop myself before doing as I always do, but I am nevertheless

able to see 'that did not come out very nice' and at times I can actually repair such a situation.

(Site manager 2)

As the quotation suggests, the site manager thought of himself as being a spontaneous person, not always being overtly concerned with his actions and statements, but still anxious to make things work. The coaching process helped him, he thought, to calm down and reflect on a situation. He was positive about the coaching programme because it provided him with tools and techniques, but also the time to step out of the flow of everyday life:

> You could express it accordingly: Now I have the feeling that I have a better opportunity to *become* a better site manager than I did not have previously. I do not think that 'now I am fully skilled' ... I may even have taken a step back since I am becoming aware how many things to actually use and explore in situations where I am used to shoot from the hip. But with a little insight and understanding, I may be able to take a step forward.
>
> (Site manager 2)

However, he also felt some pressure to use the tools and methods actively because he could now no longer 'say that I did not know how to do it'. Yet another effect from the coaching programme was the symbolic function of being a coachee. One of the site managers pointed at the coach as a door-opener that gave legitimacy to new issues previously marginalized or ignored in the industry or at the site:

> It would have been a damn stupid thing to say that you don't think it helps you, because I think it does ... [co-workers says] 'well, here comes the coach again' and then we can joke a bit about it and still there are some questions popping up around all this – like 'what do you actually talk about?'. Then you get a focus on these things and you can continue discussing with individuals or during the site meetings once a week.
>
> (Site manager 5)

All but one of the site managers – who had an unfortunate work situation, being in between two positions during the period and thus thought the programme demanded too much time – were either very positive or positive about being part of the coaching programme. One of them emphasized the need for having someone to discuss ideas with:

> I think it has been a privilege to join the programme. You get new pieces all the time; if you are in a stressful situation and you sit down to reflect, you may actually be able to handle the situation better ... As a

site manager there are few superiors you can consult in this manner; you have to appear as the leader and the motor of the workplace.

(Site manager 5)

However, the site managers expressed some concerns regarding coaching that need to be addressed when enrolling in such a programme. One of the site managers pointed at the degrees of freedom in the programme and questioned whether 'each and everyone is capable of working under such free conditions' (Site manager 1). In his view, some coachees would be able to learn more from the programme if they were given a clearer 'course plan'. Another site manager said that the responsibility lies very much with the individual and therefore coachees should be selected carefully, preferably among younger, dedicated site managers needing help and direction to carry out their work. It is also important to recognize that the need for coaching support may not lie in the areas prioritized by the company:

> From the perspective of the firm, it is always easy to prioritize the heavyweight site managers because they are the ones with the highest turnover. But it is actually the smallest sites where they are most exposed to psychological pressure ... I am familiar with the role of those travelling around between different sites, and they endure a damn stressful situation, to say the least.
>
> (Site manager 5)

Prioritizing the already effective site managers at the expense of the more marginal groups would, however, mean that the opportunities with coaching were not being fully exploited. The site manager continued: 'This type of coaching is good for people thinking this is a really tough situation and being virtually down on their knees. They are the ones with the greatest need for active reflection' (Site manager 5). Another site manager argued that there is always cost, effort and time involved in coaching, but that his experiences had been positive enough fully to endorse a broader use of the method to help fellow site managers. He also argued that the ability to delegate is one of the principal gains from the coaching programme, benefiting the industry the most:

> It takes time ... it costs time and money ... that is actually what we need to learn, to let ourselves have some time occasionally, during those eight hours we spend at work every day ... If we let our co-workers deal with these issues – they are in fact just as qualified as we [site managers] are but they come here to ask us anyway – then we would be able to look up and anticipate forthcoming events and thereby be able to create a more decent everyday work situation.
>
> (Site manager 2)

Some of the site managers also reflected on how the coaching process differed in comparison to other more conventional leadership training courses. The common argument was that the strength of coaching was, beside its strong orientation towards individual and workplace-based concerns, the repetitive mode of learning. One of the site managers argued:

> If you attend a course there are always some words and phrases sticking to your brain, but most of them end up on the shelf and then it is all gone, while here it is repeated over and over in a positive manner.
>
> (Site manager 2)

Conventional courses were then, in the eyes of the site managers, complementary to a coaching programme rather than being a full substitute.

Most of the site managers participating in the coaching programme emphasized that coaching could be part of a broader restructuring of the industry. Traditionally, the construction industry (at least in the case of Sweden) offered little support for its site managers. One of them described how he was introduced to the profession:

> 'Here, you have a bunch of blueprints, off you go! Do you know where the site is?'. 'Not really'. 'Well, there's a map in there you can take, so you can find your way. The client is at this address, you can go see him' … And then you have to fix it on your own, all that you need to do. It is often taken for granted that the site manager is capable of operating on his own, right.
>
> (Site manager 4)

Another site manager agreed with such criticism:

> You need to think in new terms. You can no longer contrive your own solutions to everything. There is this thing with the construction industry: there is an inadequate support for the managers. You become site manager relatively soon if you are a bit ambitious and then – they give you a few courses, that is true – you're on your own. They give you the calculations and blueprints and then 'build this and make some money', and as long as that works satisfactorily, then nobody really cares even though there are numerous things to improve.
>
> (Site manager 5)

He continued:

> It is of *critical importance* that construction companies become able to take these things seriously. Younger people today put into question this way of working; nobody will be working as they used to do,

I don't think so. There is a need for a completely new organization. We need to think in new terms.

(Site manager 5)

Another site manager claimed that the 'macho image' of the site manager as someone capable of enduring any conceivable work conditions and being able to solve anything on his own, needed to be abandoned once and for all. Hence, the coach needs to represent a more 'contemporary view' of leadership:

The site manager role is a bit of a stereotype; one needs to be tough and hard and so on, and I think that is exactly what we need to get away from. If there is someone [i.e. the coach] saying 'you gotta be tough, y'know' ... it doesn't work like that.

(Site manager 2)

Yet another aspect of the coaching programme was that it may help recruiting new site managers:

Today, it is hard [Swedish, *svårt*] to recruit a site manager. Above all, it is known that it is a hell of a responsibility and you're insecure regarding what role you're accepting ... If there is such a [coaching] programme, that would make recruitment easier, I'd say. You'd think of the programme as a form of support.

(Site manager 3)

Taken together, the coaching programme was positively received by all of the site managers, even though one could not fully commit to it owing to an excessive workload. The principal outcomes were more attentive and thoughtful action on the part of the site managers, and a better understanding of the objectives and roles of different actors in construction projects. In addition, coaching was regarded as complementary to more regular courses and training, and was also seen as an approach supporting site managers in their day-to-day work

The coach's view of the process

The coach, who had previously worked with individual managers but never with a cohort of six persons belonging to the same profession, thought of the programme as being interesting and a real challenge. For the coach, it was important not to pursue any specific predefined goals with the programme but to help the site managers open up to new ways of thinking about themselves, their skills and shortcomings, and their broader social and professional lives. One could say the coach endorsed a pragmatic approach to the coaching work. He defended such a position:

> I am not 'religious' in the sense of the term that they need to share my beliefs, or like what I like, but I have, to use a humanistic vocabulary, 'embraced them' and their ideas. I have tried to strengthen their identity and their beliefs at the same time as I attempted at helping render such beliefs problematic: 'What are the risks with that approach?'. Not playing the role of a mentor saying 'do it like this … '. Coaching is about improving the performance with available skills and capabilities.
>
> (The coach)

The coach, whose experience came from a leadership training programme and former doctoral studies in education science, was very clear that his role was not that of an 'expert' capable of 'telling people what to do', but merely that of an experienced and knowledgeable interlocutor. He was thus concerned about expectations on 'expert leadership' implicitly assumed by the site managers:

> They had certain expectations that they would meet some expert, a 'site manager expert' delivering answers to all kinds of questions. They realized straight away I wasn't that expert and some of them were even a bit disappointed. At the same time, I sort of 'won the battle' when they realized that my domain of expertise was virtually unknown to them and that if they would 'lost out in one area' they would gain so much more elsewhere … To make use of coaching, one should really be aware of the risks of playing the role of the expert.
>
> (The coach)

In other words, the site managers were not provided with detailed technical instructions on how to cope with various concerns and challenges, but were given the opportunity to articulate and discuss what they perceived to be their major concerns for the time being. The coach argued that this articulation is helpful for structuring their work life:

> When they sit down with me, they need to be precise and to elaborate on their thoughts. When they structure their concerns they can relate to them in a new manner. They are not *within* the process but can take a step aside and *perceive* the process. I think that is what they have noticed … It is like a calm space; they get a comfortable retreat from work when I arrive and create opportunities for reflection, in turn engendering processes of thinking.
>
> (The coach)

In summary, the coach carefully avoided being someone providing ready-made, off-the-shelf solutions to managerial problems emerging in the work of a site manager, but actively pursued a relationship with the coachees where they could sit down and relate to their own work, and their own skills and challenges.

The coach thus designed the programme to make the coachees feel comfortable and secure enough to bring their own personal concerns and address their perceived shortcomings. However, rather than just giving them credit for this, he also wanted them to think in new terms and view their work-life situation from alternative perspectives. The coaching programme was in general aimed at helping the site managers 'help themselves' through emphasizing new modes of thinking and new 'perspective-taking'.

Managing knowledge through coaching: some lessons learned

The study of the coaching programme wherein six site managers participated shows that site managers appreciated coaching as a form of systematic support for their work. The site managers pointed at their new skills in reflecting on their work practices and their overall life situation, their ability to take the role of the other and perceive things from many perspectives. Moreover, they thought the programme had certain symbolic effects enabling an extended discussion on leadership practice and project goals with the co-workers in the project. Moreover, all the site managers argued that it was of vital importance that construction industry companies actively engaged in transforming the site manager's role into a new position wherein he (or, to a lesser extent, she) would not have control and monitor all activities in detail but where responsibilities and tasks to a larger extent would be distributed between project co-workers. The site managers thus conceived of coaching as one approach among others to support site managers. In addition, they emphasized the different modes of learning provided by a coaching programme and a regular course. The coaching programme was embedded in the site managers' practical concerns rather than addressing more generally formulated challenges, and the integration of theoretical discussions, reflection and practical work provided an effective arena for not only learning new methods but also for conceiving of themselves and their role and aspirations in new terms.

The study contributed to the literature on construction management in terms of showing that coaching could be effectively used to support not only site managers but also most categories of managers in the construction industry. However, given the site managers' central position in construction projects and their exposure to workload, complex assignments and legal responsibilities, the site manager is a position benefiting from the support provided by a coaching programme. As suggested by the research reported by Djerbarni (1996) and Lingard and Francis (2004, 2006), site managers endure a significant workload, are less satisfied with the pay, work long hours, and perceive work-family conflicts in their day-to-day work. As Lingard and Francis (2006) show, perceived support from the employing company significantly reduces the risk of burnout and stress for these groups exposed to a variety of stressors. The absenteeism of site managers is a substantial cost for any construction company, and providing support that reduces stress and work–family conflicts may be a sound investment for the future. Seen in this view, coaching is not some kind of 'add-on' service to comfort managers but an important strategic human-resource and management practice whose function and role deserves further attention and research in the construction industry. It is also noteworthy that coaching demands full commitment and a willingness from the coachees to share their beliefs, thoughts, and even hopes and dreams with the coach. It may be that this specific form of support and training opportunity is not ideal for all site managers (or managers in general in the construction industry), and that some may prefer a less personal form of support. Some site managers in the study also suggested that coaching would be of most value to younger site managers because more experienced site managers may fail to question their own work procedures actively. However, one of the site managers, who was over 40 but rather inexperienced as a site manager, claimed he benefited greatly from the programme, but also admitted he might be close to the age where it may no longer be useful to join the programme.

The study also contributes to the literature on middle managers, suggesting that middle managers are stuck between top management's strategies and enacted objectives, and the demands and expectations from co-workers and the practical concerns of everyday work. It also shows that there are tools and methods actively supporting middle managers to improve their leadership skills and their ability to delegate responsibilities to co-workers. Since knowledge tends to be what Gabriel Szulanski calls 'sticky' (Szulanski, 1996; Von Hippel, 1998; Szulanski and Cappetta, 2003), that is, complicated to transfer and share because it in most cases comprises both explicit and propositional knowledge and tacit knowledge resisting articulation and formal representation, the communication skills of practising managers is always of great importance. Following Szulanski (1996), knowledge does not flow freely in organizations, but must be carefully managed and shared; here skills appropriated from coaching activities may make a substantial contribution to the day-to-day work. As one of the site managers suggested, being able to 'take on the role of the other' and to perceive the assignment from many angles helped him improve his communication and make the co-workers collaborate more effectively. The site manager's and the middle manager's role are not then defined once and for all and do not need to remain static, but can be actively transformed when new opportunities emerge. Thus, the somewhat bleak image of the middle manager needs to be put into question.

Finally, the study is a contribution to the coaching literature in terms of giving voice not only to professional coaches, advocating their tools and methods, but also to the coachees, the practising site managers in construction projects. While virtually all coaching literature is either presented as checklists and general statements on 'how to coach' or describes coaching programmes in very general terms and with little empirical illustrations, this study allows for a more empirically oriented approach to coaching wherein the directs benefits and concerns are addressed. While coaching is potentially a powerful method of supporting leaders and managers and developing human resources and leadership skills, the coaching literature remains largely preoccupied with presenting and introducing coaching. Consequently, there are few studies on how coaching is actually received by the managers themselves. In the future, we will hopefully witness the growth of literature on empirically oriented coaching that is less concerned to legitimize itself *vis-à-vis* other leadership development methods.

Theoretical implications: coaching as second-order observations

In this section, some theoretical implications for the organization-theory and knowledge-management literature will be addressed. In discussing the value of executive coaching in the construction industry and knowledge-intensive work, the work of the German sociologist Niklas Luhmann and, more specifically, his concepts of first- and second-order observations will be examined in some detail.

Niklas Luhmann is today widely recognized as one of the most influential contemporary social thinkers in the domain of functionalist sociology. His oeuvre is massive and, to date, only a limited amount of his work has been translated into English. Luhmann's work is wide in scope and addresses legal theory, sociology, complexity theory, art and aesthetics, media studies, and even culturally embedded concepts such as love and passion. Even though Luhmann's work is extensive and complicated to cover fully, his work is also consistent and systematic. He regarded himself a follower of Talcot Parsons (1951), defending a formal and functionalist sociology, but integrated, *inter alia*, the works of the Chilean biologists Herbert Maturana and Francesco Varela (1980, 1992) on what they call *autopoesis* and *autopoetic system* into his social theory. The English mathematician George Spencer-Brown is another influential thinker carefully attended to in Luhmann's work. Recently, Luhmann has been increasingly recognized in organization theory and management studies (Bakken and Hernes, 2003; Nassehi, 2005; Siedl, 2005) and has been used in a range of studies of organization procedures (Hernes and Bakken, 2004; Czarniawska, 2005; Sundgren and Styhre, 2006). However, the secondary literature on Luhmann frequently speaks of his work as being 'abstract' and 'complicated' and Luhmann (2000a) himself insisted on examining society a priori. For instance, Bakken and Hernes (2003: 9) say that 'Luhmann's work, in all its richness, is complex and appears inaccessible to many students of organization'. Luhmann's thinking is perhaps most easily apprehended in the writing wherein he examines idiosyncratic social systems such as media (Luhmann, 2000b) or art (Luhmann, 2000a). On the other hand, Luhmann's alleged 'difficulty' must not be overrated; his work is opening up for many useful insights into organizations and managerial practice (see e.g. Czarniawska's, 2005, analysis of paradoxes in organizations).

It is outside of the scope of this section to account fully for Luhmann's work but some of the elementary ideas will be addressed. First, Luhmann speaks of society as the totality of interrelated social systems. For Luhmann (1982: 70), one may speak of a 'social system' whenever 'the actions of several persons are meaningfully interrelated and are thus, in their interconnectedness, marked off from an environment'. The next distinction is that between the system and the environment that the system is relating to but not being determined by. Luhmann (1990: 85) also strongly distinguishes between *living systems,* based on the 'life processes' of a biological organism, *psychic systems* based on 'consciousness', and *social systems* based on 'communication'. These different systems are interrelated but remain strictly separated qua analytical categories. Moreover, all social systems are autopoetic systems, that is, they 'are the products of their own operations' (Luhmann, 2002: 102). Luhmann continues:

> They have properties such as dynamic stability and operational closure. They are not goal-oriented systems. They maintain their autopoetic organization of self-reproduction as long as it is possible to do so. Their problem is to find operations that can be connected to the present state of the system.
>
> (Luhmann, 2002: 102)

The autopoetic reproduction of systems is different in living, psychic and social systems. In social systems it is the 'capacity for action' (Luhmann, 1995: 372) that determines the viability of the system. Moreover, since social systems are based on communication and nothing else, a point forcefully repeated by Luhmann throughout his work, communication produces concepts such as trust, meaning, and reflection, enabling the continuation of communication. Communication is conceived of as the integration of *information, utterance* and *understanding* in one single process. Seen in this way, communication is always capable of failing; a word may be misunderstood, information may no longer be useful, an utterance may be faulty. There is then no inherent rationality in communication, Luhmann argues. Communication cannot promise anything but further communication – it is autopoetic. Concepts like reflection and trust are for Luhmann functional mechanisms safeguarding the autopoetic reproduction of social systems: 'Reflection may ... be defined as the process through which a system establishes a relationship with itself ... reflection is a form of participation' (Luhmann, 1982: 327). Elsewhere he argues: 'Reflexive mechanisms extend the potential for complexity of the society and thus the prospects of survival for the social system in which they are instituted' (Luhmann, 1979: 66).

The concept of second-order observations forms part of this ongoing communication autopoetically reproducing social systems. For Luhmann, in the communication of social systems, one must always separate between the observations of actual occurrences and events qua primary qualities, that is, *first-order observations*, and *second-order observations*, that is, the observations of first-order observations. This distinction is probably most easily illustrated with reference to Luhmann's analysis of the social system of art and the distinction between *value* and *price*. In the market for art, actors praise or appreciate the value of a particular painting and are willing to pay a specific price for the work of art. On the level of second-order observations, art dealers and art market analysts can do away with the concept of value or 'use-value' but are primarily concerned about the price and, more specifically, changes and fluctuations in prices. In social systems, first- and second-order observations are mutually dependent because, without the ability to make second-order observations, the art market would not function properly. Luhmann (2000a) explains:

> Without market-dependent prices, there can be no second-order obser-vations and thus (as socialist state planning learned the hard way) no specific economic rationality. That is why economic theory must distinguish values and prices, depending on whether it observes an observer of the first or of the second order.
>
> (Luhmann, 2000a: 64)

Not only do the markets operate effectively through blending first- and second-order mechanisms, but all social systems are dependent on this separation. For instance, art criticism, embedded in another form (to use Bourdieu's term, essentially incompatible with Luhmann's sociology) of cultural capital, is drawing on the skills of making legitimate second-order observations:

> Looking at a painting, listening to a piece of music, or simply identifying a work of art (as opposed to another object) from a first-order observer position does not yet imply a capacity for judging the work. The naked eye does not recognize artistic quality.
>
> (Luhmann, 2000a: 80)

In order to make second-order observations, the actor (e.g. the art critic) must be familiar with other utterances on the level of second-order observations. The reception of a new work then never occurs in a social vacuum but is always already located in a domain constituted by second-order observations. The concepts of first- and second-order observations are also applicable in the analysis of meaning. Luhmann (2000b: 93) says, with reference to classic sociology, that humans are self-observing and become what they are through shifting between first- and second-order observations:

> Individuals are self-observers. They become individuals by observing their own observations. Today, they are no longer defined by birth, by social origin, or by characteristics that distinguish them from other individuals. Whether baptized or not, they are no longer 'souls' in the sense of indivisible substances that guarantee an eternal life. One might argue with Simmel, Mead, or Sartre that they acquire their identity through the gaze of others, but only on conditions that they observe that they are being observed.
>
> (Luhmann, 2000a: 93)

In summary, Luhmann's systems theory suggests that, in order to function properly, communication must constantly mingle first- and second-order observations. Still, concepts such as value and price, primary impressions of an artwork and legitimate commentaries on it must remain separated. In terms of management coaching, this distinction is useful because it underlines the coach's role to promote second-order observations actively, which in turn may influence action and perception at the first level.

Speaking of coaching in Luhmannian terms would be to emphasize how communication and reflection is produced through new forms of communication, namely communication introduced by the coach. In more mundane terms, rather than continuing communicating along well-known routes, speaking of everyday matters such as, for example, effectiveness, performance or conflicts between co-workers, new information and new forms of utterances may be brought into the communication by the coach. New phrases, new perspectives, new theoretical models and so forth pave the way for communication in new directions. In addition, the coach's principal role in this perspective becomes to underline the need for being capable of making second-order observations in day-to-day work. Next, we will examine the work of the practising site manager in the construction industry.

As pointed out by commentators such as Gray (2006), the literature on coaching is not very articulated in terms of its theoretical and epistemological assumptions. Instead, it is concerned more with praising the benefits of coaching and pointing out its potential effects on management practice. In addition, few empirical studies provide any details of what happens in coaching programmes or give voice to participating managers. Rather than just providing checklists and 'how to coach' advice, the coaching study suggests that, despite early concerns and the challenge to abandon preconceived ideas, the coachees appreciated the coaching programme. Of particular interest for the site managers was that they were given an interlocutor with whom they could articulate (see e.g. Patriotta, 2003) and 'emplot' – in Hayden White's (1987) sense of the term – their concerns, that is, rendering concerns meaningful and manageable through imposing an intelligible structure, a 'story-line' on the event. This articulation helped site managers take a step back and to perceive their own work-life situation, a first move towards a more comprehensive re-evaluation of individual work practices.

Drawing on the theoretical framework formulated by Niklas Luhmann, one may argue that executive coaching is providing a systematic way of providing opportunities for second-order observations. In Luhmann's systems theory of society, equally psychic systems based on consciousness and social systems based on communication are dependent on the possibility for both first- and second-order observations. Humans are constituted as coherent individual subjects through continuous and ongoing observations of the self and others' observations of the self. Social systems, constituted by communication, demand mechanisms enabling second-order observations to make communication function properly. In the case of coaching, the coachees – in our case site managers – bring their first-order observations to the coaching sessions; site managers report recent events, occurrences, actual and potential conflicts with co-workers or sub-contractors, and concerns for forthcoming events may be addressed. On the basis of these first-order observations, the coach and the coachee jointly sort out why the site manager is relating to a particular problem in a certain way, how he has responded to the problem, etc. The coach helps the coachee to 'step out' of the situation and account for his own actions or non-actions, behaviours, responses and other relevant reactions to the particular event. During the process, the coach interrogates the coachee and poses specific questions, helping him to create a meaningful experience out of the event. A meaningful experience here includes discussions about personal traits and preferences, past experiences from similar events, and other individual aspects, but also the use of and reference to adequate theoretical models and theories capable of shedding some light on the object of discussion. Speaking in Luhmann's terms, the coach increases the informational content of the communication – itself constituted by the integration of information, utterance, and understanding – but also enables understanding through the articulation and emplotment of a specific learning. For instance, when addressing a conflict between a site manager and a foreman, the coach helps the coachee to relate to the conflict not only through a discussion about previous conflicts and the site manager's personal way of dealing with conflicts, but also through presenting theories and models of what conflicts are, and how they evolve over time, and other relevant characteristics of conflicts. The site manager becomes defamiliarized from his initial first-order observations and can relate to the conflict in a more meaningful manner, perhaps even being capable of changing his own behaviour.

Examining coaching as a blend of first- and second-order observations provides the coaching practice with a more comprehensive theoretical framework from which coaching can be examined. Rather than advocating coaching on the basis of personal experiences and preferences, coaching becomes located within a theoretical framework explaining why coaching is potentially a powerful leadership development practice. The study also contributes to the coaching literature in terms of giving voice not only to executive coaches but also to the coachees, a group that is strangely absent in the coaching literature. Most of the literature speaks of the outcomes from coaching but little is said about the process taking place in any coaching programme. Moreover, the chapter contributes to the literature on site management in construction industry and the more general project-management literature by suggesting that site managers and project leaders are exposed to a work-life situation characterized by ambiguities, a substantial workload and stress. The chapter also suggests that coaching, under specific conditions and with the requisite level of commitment from the participants, is a useful approach to support site managers in developing their leadership skills and dealing with their everyday work.

Summary and conclusion

Previous studies suggest that site managers in the construction industry are exposed to a work situation characterized by a substantial workload, multiple leadership roles, legal responsibilities, and a certain degree of isolation from other site managers and superior managers. In order to support the site manager in his or her work, new tools and methods need to be developed. This chapter reports a study of a coaching programme in which six site managers participated, and suggests that coaching provides an arena wherein the practical concerns of the site manager are addressed and combined with theoretical and practical training and systematic reflection. The study shows that site managers appreciated the coaching support and were able to identify several immediate practical effects for themselves, their co-workers and the construction project in general. The chapter thus contributes to the literature through providing evidence of the value of coaching in the construction industry, and through presenting one of very few studies of coaching that actively give voice to the coachee rather than solely presenting coaching as a legitimate leadership development method.

3 Materiality, aesthetics, and social and cultural capital in architects' work

Introduction

In this chapter, the system of knowledge mobilized in architects' work is examined. Since the work of the site manager considered in the previous chapter is rather different from the work assignments facing the practising architect, this chapter uses a complementary theoretical framework to address the skills and capabilities involved in architects' work. In the discussion of this work, a distinction is made between an internal and an external perspective. The internal perspective emphasizes the day-to-day communication among peers centred on models, images, blueprints and other artefacts serving as boundary-objects in architectural work. The external perspective will point at the value of credit and reputation in the field of architecture outside of the focal firm. Architects essentially pursue, similar to many other professional groups and in the performing arts, what has been called 'portfolio careers'. That is, the entire career trajectory of a practising architect may be laid out as a series of concurrent or succeeding projects that collectively constitute an individual's career. Every architect can tell stories of what projects they participated in and what competitions they have joined or may even have won; their working life is, just like in the other fields of the construction industry, essentially 'projectified'. In the analysis of the internal work processes, architects' work will be examined in analogy with scientific laboratory work. The relative lack of detailed studies such as ethnographies of architectural practice and the abundance of similar studies of laboratory research motivates such a comparative study. Studies in science and technology have traditionally provided a great number of highly detailed ethnographies that are arguably helpful when understanding what architects do in their day-to-day work life and when examining what sources of knowledge they use in their professions. Expressed differently, there is a certain *morphology* shared between professions notwithstanding their idiosyncrasies and specializations.

Architects' work: an overview

Studying aesthetic knowledge at work

In the wake of recent interest in the various forms of knowledge mobilized and used in organizations, a great variety of professions and occupations, as diverse as restaurant cooks (Fine, 1996), salesmen and fast-food restaurant workers (Leidner, 1993), financial analysts and stockbrokers (McDowell, 1997; Hassoun, 2005), flight stewardesses (Tyler and Taylor, 1998), technicians (Orr, 1996), corporate managers (Jackall, 1988), laboratory researchers (Lynch, 1985; Owen-Smith, 2001), or even more shady 'occupations', such as drug-dealers (Adler, 1985) or sex workers (Sanders, 2004), to name a few, have been studied. In many cases, knowledge is regarded as the capacity to interpret and manipulate symbols and texts, for instance, in the case of computer programmers or market analysts in the financial sector of the economy. In other cases, it is the ability to integrate analytical skills with other human faculties, for instance, perceptual and tactile capacities, that constitute idiosyncratic expertise. For instance, physicians, such as general practitioners, are trained in bridging theoretical and explanatory frameworks and individual observations made in the diagnosis of the patient (Foucault, 1973; Mol, 2002; Clarke, Mamo, Fishman *et al.*, 2003). In such professions or occupations, knowledge work is embodied in terms of being the outcome from a close proximity between the cognitive and embodied capacities of the knowledge worker. In general, there is a shortage of studies in the field of knowledge management wherein professions or occupations blend theoretical frameworks and impressions through the senses.

Architects are one such profession relying on the integration of cognitive, perceptual, tactile, communicative and even aesthetic capacities and skills. Architects influence society in terms of designing buildings and civil engineering projects that last for significant periods of time, in many cases for centuries. Even though there is a body of texts examining the social function of architecture (see e.g. Le Corbusier, 1946; Venturi, Brown and Izenour, 1977; Koolhaas, 1978; Saint, 1983; Trieb, 1996; Benjamin, 1999; Grosz, 2001; Obrist and Koolhaas, 2001) and a general interest for the aesthetics of organization (Guillén, 1997; Tyler and Taylor, 1998; Strati, 1999; Linstead and Höpfl, 2000; Hancock and Tyler, 2007; Warren, 2008; see Chapter 5 of this volume for an extended discussion) and in society at large (Julier, 2000), architects' work is rarely subject to analysis in management studies. While architects are in general not responsible for the actual production of the building, they provide the scripts (i.e. the blueprints) from which the building is constructed, and they specify the material and aesthetic features of the building. However, Kwinter (2001: 14) warns that one should not limit the concept of 'architectural substance' to building materials and 'the geometrical volumes they engender and enclose', but to examine the 'systems of legitimation' in which the building is located, designed and finally constructed. Kwinter (2001: 15) continues: 'Architecture's proper and primary function, it could be said – at least in the modern era – is the instrumental application of mastery, not only to an external, nonhuman nature, but to a *human* – social, psychological – nature as well'. Kwinter (2001) thus suggests that buildings are not only material, but also aesthetic and social entities and need to be examined as such. For instance, the widely renowned architect Robert Venture says: 'The essential element of architecture for our time is no longer space, it is no longer abstract form in industrial drag; the essential architectural element is iconography' (cited in Obrist and Koolhaas, 2001: 593). Buildings are inherently changing and socially constructed entities that are informed by social interests and political concerns (Brand, 1994); 'We shape our buildings, and afterwards our buildings shape us', Winston Churchill once remarked (cited in Brand, 1994: 3). Speaking with Kwinter's (2001) sophisticated vocabulary, any work of architecture is a *technical object*, an entity surrounded by a complex of 'habits, methods, gestures, or practices that are not attributes of the object but nonetheless characterize its mode of existence – they relay and generalize habits, methods, and practices to their levels of the system' (Kwinter, 2001: 21); a technical object is what is always already inherently social, yet material and aesthetic, and thus needs to be examined as such. Studying architecture work is therefore to bring together a range of social resources and capabilities into one single integrated analysis. Architecture work operates in many dimensions and on many layers of meaning: the layer of materiality, the layer of functionality and social practice, the layer of the aesthetic, and the layer of politics, to name a few. Architecture work is a social practice contributing to what Henri Lefebvre (1991) calls 'the production of space', the active structuring and inscription of social spaces, and is therefore of necessity a multivariate social process touching upon many dimensions of sociality.

Studies of architects' work

To date, there is little research on architects' work in management studies and, of the few studies that exist (Blau and McKinley, 1979; Winch and Schneider, 1993; Kamara, Augenbroe and Carillo, 2002; Pinnington and Morris, 2002; Ivory, 2004), most primarily examine the formal organization of architects' work and not the practice *per se*. Ankrah and Langford (2005) examine the organization culture of architects' offices and construction companies, and characterize the architectural practice accordingly:

> Architectural practices are largely informal organizations in which control and coordination are achieved through empathy between organizational members and through direct personal contact, and this is essentially because most of these practices are small. There is

decentralization in decision-making with everyone encouraged to think and contribute to problem-solving, although the managing director (usually the founder or a founding partner) plays a pivotal role in coordination. These organizations employ highly trained and skilled individuals who have a high tolerance of ambiguity. Their sense of their own importance creates in them a need for recognition and a desire to impose their identities on the organization. The firms in this business recognize their employees' importance and, accordingly, acknowledge and reward their individual efforts and performance, although to some, not enough.

(Ankrah and Langford, 2005: 601)

Symes, Eley and Seidel (1995: 31–8), in their survey of 610 British architects, found that skills in design and creativity were important, communication skills are valued, architects are regarded as being in possession of 'special knowledge', 'flexibility and personal style' are necessary, and that marketing expertise is demanded.

Architects are working in environments where decision-making is decentralized, personal contacts are of central importance, and where there is a 'high tolerance of ambiguity'. They are thus professional knowledge-workers engaging in advanced collaborations with a variety of stakeholders. However, Pinnington and Morris (2002: 196) argue that architecture '[m]ay be said to be distinctive from other professions if we consider its historical development and specific relationship to its clients'. However, as a professional group, architects have witnessed a weakened authority over clients and contractors as new forms of contracting have been established. As a consequence, architects have sought to '[p]reserve a narrowed professional jurisdiction' over the aesthetic features of the building (Pinnington and Morris, 2002: 190). Moreover, Blau and McKinley (1979: 216–17) report that innovative architecture firms emphasize creativity and the ability to deliver 'unique, aesthetically or technically notable projects' more than financial performance. Such a view is supported by Pinnington and Morris (2002: 197), suggesting that '[i]t is evident that architecture firms are not simply concerned with profit maximization, but with aesthetic and public-good outcomes that are likely to influence their values and organizational decisions' (see also Cuff, 1991: 219). However, the studies of Pinnington and Morris (2002), and Blau and McKinley (1979) say little about how architectural work-practice evolves and thus leave architects' actual work very much as a black box.

Instead, one may resort to a number of detailed studies of architecture as practice (Blau, 1984; Gutman, 1988; Cuff, 1991) offering more detailed analyses of the everyday work-life of architects. Cuff (1991) portrays architectural work as a trade riddled by ambiguities and emergent properties demanding a continuous re-negotiating of the design process. As a consequence, architects need to conceive of their work, Cuff contends, as a *social process* bringing together a variety of stakeholders: 'The most overarching observation is that the production of places is a social process. That is, a very basic task of architectural work is to collect all participants, both in the office and out, to develop a manner of working with them and to interact with them in order to create a design solution' (Cuff, 1991: 245). At times, this process is effectively managing to take into account a variety of interests while in other cases the architects fail to accomplish this. For instance, Ofori and Kien (2004), studying 'green architecture', i.e. the influence of ecological thinking in the profession, among the Singapore community of architects found that '[a] discrepancy between what architects claim to be convinced about, and knowledgeable in, and their commitment and practices; architects seem to be unable to translate their environmental awareness and knowledge into appropriate design decisions' (Ofori and Kien, 2004: 34). The Singapore architects thus embraced ecological ideologies but essentially failed to produce buildings taking into account such concerns.

One concern with these previous studies of architectural work is the lack of a point of reference; architect work is essentially examined as a discipline separate from other forms of knowledge-intensive work. However, some studies offer a more detailed account of how architects work in day-to-day settings. A series of studies examine, for instance, how architects use visual representations in their work (Ewenstein and Whyte, 2007a, 2007b; Yaneva, 2005; Unwin, 2007; Whyte, Ewenstein, Hales and Tidd, 2007). Ewenstein and Whyte (2007b: 81), presenting an ethnography of a British Architect bureau, argue that visual representations play an intermediary role between 'knowledge and knowing' in terms of communicating meaning symbolically:

> [T]hey [visual representations] communicate meaning symbolically. This helps to articulate, exchange and understand design ideas. Second, they are manifest in the practice as material entities, often physical artefacts, with which practitioners can interact as they generate knowledge individually or collectively. The communicative and interactive properties of visual representations constitute them as central elements of knowledge work.
>
> (Ewenstein and Whyte, 2007b: 82)

Seen in this view, visual representations are serving to temporally fix collective meaning. Whyte, Ewenstein, Hales and Tidd, (2007) suggest visual representations can also be 'frozen' or 'fluid'. They are fluid or 'mobilized' whenever '[c]omment, input or modification is required of certain actors in the design process' and they are frozen when visual materials are '[c]haracterized by greater certainty' (Whyte, Ewenstein, Hales and Tidd, 2007: 22). As the design process proceeds, more and more features of the building are frozen as decisions are made. For Whyte, Ewenstein, Hales and Tidd (2007) sketching is the process in architecture work wherein problems are defined and solved and, therefore, it is central to professional identity:

> In the architectural design firm visual materials are treated as fluid in the process of defining design problems and exploring appropriate solutions. The work of both individual designer and design teams is characterized by copious sketching. Through sketching, red-lining, and generally marking up representations design problems are identified and defined, and corresponding solutions are tested on paper and on screen. The objects that allow such knowledge work to proceed are fluid materials in which the status of visual representations is often provisional and in flux.
>
> (Whyte, Ewenstein, Hales and Tidd, 2007)

In addition, the use of fluid and frozen visual representations is a means for maintaining control over the research process and for mediating power relations. Whyte, Ewenstein, Hales and Tidd (2007) explicate this position:

> Practitioners can mediate power relations among themselves and other stakeholders by managing what is and is not shown, how it is shown and when. Frozen materials can more easily steer meaning and project narrative and thereby influence the kinds of understandings different actors develop. They can become a tool for the exercise of power.
>
> (Whyte, Ewenstein, Hales and Tidd, 2007: 23)

At the same time as the use of visual representations is part of the individual architect's professional skill and identity, as well as being part of the communal order of the community of architects, the constitution of meaning in the sketch-work is always at stake. Therefore, meaningful practices must always be carefully managed by the architects, both when working inside the bureau and when communicating with other stakeholders. Whyte, Ewenstein, Hales and Tidd (2007) employ a musicological metaphor to capture the fragility of the process:

> In experimental jazz, sounds may be on the edge of cacophony, in danger of disintegrating into sheer noise, from which the poorer ensembles are unable to retrieve a meaningful experience. In the same way, without care, meaning may simply evaporate in design work. An inexperienced or inattentive design team could easily lose track of what was going on and be left only with sheer information. That that is not going to happen is a testament to the designers' collective skill.
>
> (Whyte, Ewenstein, Hales and Tidd 2007: 23)

Nicolini (2007) argues convincingly that the use of visual representations must always be examined as being '[p]ut to use in the context of complex and "messy" practice' (Nicolini, 2007: 578), that is, visual representations are always achieving what he calls their 'performative power' when used in conjunction with other elements of a specific social practice. Therefore, visual representations *per se* are the carriers of meaning but meaning evolves in a broader network of relations. In addition, he points out, visual representations are always embedded in an individual or collective set of competencies that often remain hidden for both participant and observer (Nicolini, 2007):

> In order to function, so to speak, visual artefacts require a certain amount of work that often remains hidden, what could be called the 'non-visual work' necessary for making visual artefacts work. Put differently, while it is undeniable that a picture is often worth 'a thousand words', it is also true that pictures and words work together according to a subtle and often unnoticed division of labour.
>
> (Nicolini, 2007: 578)

Expressed differently, everyday work with visual representations demands a series of 'non-visual' competencies and skills that may remain more or less tacit and therefore need to be learned.

One source of concern in architect work, potentially undermining collective sense-making, is the need for being capable of continuously shifting perspective when working with visual representations. Yaneva (2005) presents an ethnography of architect work, and claims that one of the principal skills of the experienced architect is to be capable of 'scaling up and down', that is, to operate on the basis of images and models of buildings while at the same time being capable of conceiving how the actual building will appear in real life. Architects are trained to master the skill of altering perspective between what may be called the virtual (images and models) and the 'not-yet-actual' but imagined building (the future building). Given the use of many complementing models, the final building is therefore never present in a single state or model, but is instead a 'multiple object', a 'composition of many elements; a "multiverse" instead of a "universe"' (Yaneva, 2005: 871). Moreover, the models and images [e.g. computer-generated, three-dimensional (3D) images of the building] often travel outside of the architects' office to 'gain powerful allies among clients, sponsors, and future users, community groups and city planning commissions' (Yaneva, 2005: 889, note 15). The comments, concerns and articulated expectations from these groups are taken into account in the design process. Seen in this view, the models do not only incorporate technical concerns but also embody a range of viewpoints and opinions. Expressed differently, the model is serving as a 'boundary object' (Star and Griesemer, 1989), an artefact capable of providing a shared ground for conversations and collaborations across heterogeneous groups and communities, representing diverging interests. In their study of the work of the internationally renowned American architect,

Frank O. Gehry, and the use of new '3-D coding', Boland, Lyytinen and Yoo (2007) found that the traditional relations between actors in the construction process were overturned and replaced by new and, in many cases, productive and meaningful relations. The actors knew how to 'encode, classify, and transform ideas with the 2-D code shared by all participants' (Boland, Lyytinen and Yoo, 2007: 639) but in the 3-D coding the 'traditional contract language' and the associated principles of 'loose couplings' between actors no longer applied. Therefore, the actors had to collaborate more closely in the process. For instance, one of the chief operating officers in one of the participating construction firms reported that in his 20 years of working in the industry, he had spent only an estimated eight hours in architects' offices. In the Gehry project, he was expected to work very closely with the architects. Boland, Lyytinen and Yoo (2007) conclude that the media of representation strongly affects the relationship between actors in the construction industry.

Yaneva (2005) also points at a number of terms in use in architecture practice. First, they use the term 'concept' to denote the main idea of a building, 'taken in its relationships with the client demand, the city, the urban fabric, and the broader social, political and cultural context' (Yaneva, 2005: 891, note 31). The concept is thus something similar to a 'script' for the forthcoming building, a set of specifications to take into account. Second, architects use the term 'programme' to denote the 'content of the building', that is, 'the internal distribution of spaces according to functional needs, general scope and insertion in reality' (Yaneva, 2005: 891, note 33). While the concept sets the boundaries – practically, functionally, financially, politically – for what may be accomplished, the programme is the actual 'fleshing out' of the building its actual design. The models used to communicate internally and externally are designed to integrate both the concept and the programme. Accomplishing effective communication centred on models is one of the principal skills in any architects' office and being able to review, comment on, and discuss models and images of emerging buildings is a key activity in architecture work. Ewenstein and Whyte (2007b) refer to such capabilities in terms of 'aesthetic knowledge', that is, knowledge that is embodied and derives from practitioners' ability to use their visual, audible, olfactory and tactile skills in their day-to-day work. In their study of architect work, Ewenstein and Whyte (2007b) found that aesthetic knowledge plays a central role and that this aesthetic knowledge in many cases is tacit and complicated to articulate; it is, in short, not 'conceptual' but 'intuitive'. When architects collaborate, they are continually exchanging opinions and passing remarks on models and sketches and how the designed spaces may be 'experienced' by the end-users. The 'essence of this subject is "aesthetic"', Ewenstein and Whyte (2007b: 701) say; it is based on the sensitivities of the architects and their ability to articulate concerns or suggestions for changes. In both Yaneva's (2005) and Ewenstein and Whyte's (2007b) studies, architect work emerges as a social practice operating on the border between the virtual and the actual, navigating between existing models, sketches and images, and forthcoming actual buildings. Being able to bridge what is and what will become in the day-to-day work is part of the aesthetic knowledge being acquired and nourished in architects' training and practice. Skilled architects have developed an individual and collective capacity to perceive social space as both a virtual and an actual resource.

Architectural and scientific work: a shared morphology

In this chapter, a study of a Swedish architecture firm is reported. Rather than taking architect work for granted and treating it *en bloc* as a ready-made set of unproblematic practices and routines, the chapter draws on science and technology studies (STS) literature examining scientific work as a social practice and as a social accomplishment. Thus, architects' practice will be examined as that which shares certain morphology, a form, with scientific laboratory work. The principal argument is that, while scientists construct 'nature' (or rather specimens of nature) in their laboratory (Latour and Woolgar, 1979; Knorr Cetina, 1983; Lynch, 1985; Latour, 1988; Traweek, 1988; Fujimura, 1996; Rheinberger, 1997), architects engage in a similar *materialized semiosis* (to use Haraway's, 2000, formulation) in their work. Scientists produce theories that are neither wholly abstract, nor material, but both simultaneously; theories are *embodying* material experiments and manipulation of natural specimens constructed in laboratories. Such theories then influence social reality in various ways. Architects, on their side, also operate on the border between the abstract and theoretical, and the material and practical. Similar to scientists' theories and theoretical explanations, the blueprint (here regarded as the final and ultimate outcome from the entire architectural work process prior to the actual construction process) is embodying both abstract and symbolic, and material and practical concerns. Scientists negotiate with one another on how to perceive nature and social reality through the mobilization of a variety of heterogeneous resources; architects shape social reality through combining various resources at hand. Scientists use theories, laboratory equipment, genetically modified animals, standardized methods for exploring and interpreting matter, and forms of storytelling and rhetoric to convince themselves, peers, sponsors and the general public of the need for their efforts. Architects read and inspect programmes, specifications and sketches, visualize, talk to one another, and communicate with stakeholders, such as clients and end-users, in their day-to-day work. Seen in this way, laboratory scientists' and architects' everyday work is equally composed of a series of activities wherein the abstract and the concrete, the theoretical and conceptual and the material, co-mingle and cross one another without being either fully integrated or separated. Both scientists and architects operate in the borderland between the abstract and the concrete.

Aligning the symbolic and the material in scientific practice

When examining the shared morphology between scientific practice and architect work, there are at least four traits that are shared between the two types of knowledge work. First, both categories of experts are operating in the intersection between the abstract, symbolic and theoretical, and the material and practical. A scientist's theory is a formal, often written statement, capturing some relationship between various components in a scientific experiment. Theory is here a representation of what actually occurs under determinate conditions when nature is manipulated; theory is thus neither abstract nor concrete, it is a synthesis of both. Similarly, the architect's final blueprint is not a mere fabrication on the basis of aesthetic preferences and beliefs, but is a document that embodies a range of specifications, expectations and opinions among a variety of actors, for instance, clients, future end-users, the team of architects and so forth. Similar to the scientists' contribution, the theory, the blueprint is neither abstract nor yet material: it is what *anticipates* or *represents* a building that eventually will be produced. Second, both scientists and architects are dependent on their ability to communicate and discuss their work with peers, both internally at the specific workplace, and also nationally or internationally. Theories and architectural solutions to problems are in many cases the outcome from continuous and longstanding debates and even controversies on how to perceive and interpret a specific condition (Ivory, 2004). Third, for scientists and architects alike, the principal influence for new thinking and new activities and solutions to problems comes from outside of the specific workplace or laboratory. Scientists may 'race' to explore a new field – see e.g. Fujimura's (1996) study of proto-oncogene research in the 1980s and Rabinow's (1996) ethnography of a biotechnology company – and then become preoccupied with sorting out their own findings or not even having time to await the review process. However, in most cases, scientists depend on peer review, recognition of their work and external influences to determine what is going on in the field. Architects, on their side, get their recent works reviewed and examined in publications specializing in new architecture. New thinking is often the outcome from consulting what is happening in the international field of architecture. This brings us to the fourth and final similarity between scientific work and architectural work, that of the emphasis on peer recognition or reputation. Both scientists and architects belong to professional groups wherein credibility and status within the professional community is primarily generated by the recognition of peers, not external stakeholders. These four aspects will be examined in greater detail.

Bridging the symbolic and the material

In a range of studies of laboratory practices, it is suggested that researchers do not intervene in nature, as in the use of the term in common-sense thinking. Nature 'as such' would be too fuzzy, too unpredictable and too complicated to manipulate to allow for any effective research process. Instead, scientists invent new forms of nature in the laboratory, an ersatz nature that is more 'predictable' and 'more manageable'. For instance, Joan Fujimura (1996), studying the emergence and establishment of proto-oncogene cancer research, writes:

> Scientists 'create' nature in laboratories just as they create 'intelligence' in computers. They create 'nature' along the lines of particular com-mitments and with particular constraints, just as computer scientists create computer technologies along the lines of particular commitments and with particular constraints. Thus, the boundary between science and technology is blurred. To bring the theoretical discussion back to the case of inbred mice, geneticists employed experimental technologies and protocols to create novel technoscientific objects. These objects *are* nature, at least as nature is described in scientific narratives.
>
> (Fujimura, 1996: 31)

The 'discovery' of the first so-called oncogenes, genes that were claimed to cause cancer, led to the development of a 'new nature' in terms of genetically modified animals such as, for instance, the trademarked OncoMouse, the first patented animal that was promised by its developer to develop cancer 'within weeks' (Haraway, 1997). The 'nature' in the laboratories is thus not independent of the various technologies, standardized methods and the procedures of the laboratory; instead, as Fujimura (1996) emphasizes, nature is the outcome from the establishment of such 'packages' of interrelated tools and methods. 'Scientific activity is not "about nature", it is a fierce fight to *construct* reality', Latour and Woolgar (1979: 243) say. In the laboratory, there is no longer any determinate line of demarcation between abstract, conceptual and theoretical resources – the entire cartography of a scientific programme – and the material and practical effects and outcomes. The abstract and the concrete become entangled and mixed; theories are embodiments of material experiments but they are not material themselves. The same goes for entire laboratories; they are, as Knorr Cetina (1983: 121) emphasizes 'materializations of earlier scientific selections'. For instance, the construct of a 'gene', today more stabilized, inscribed and rendered the status of being a 'fact', was originally an abstract concept. Gradually, the concept of the gene becomes useful because it was surrounded by various practices and theoretical elaborations. Fujimura exemplifies:

> Biological representations and objects are the products of collective (and sometimes conflicting) arrangements that change over time. 'Gene', for

example, is a concept that has had a different meaning dependent on its situated practice and use. Currently, a new set of commitments and practices have created new technical definitions of genes and genetics in molecular terms. 'Gene', then, is not an abstract concept when situated in the work of biologists, their research tools, the purposes and themes that framed their work, and the uses to which the work is applied.

(Fujimura, 1996: 65)

Speaking of architect work, architects are, as we will see, using a number of tools and practices to transform abstract ideas and pre-figural aesthetic ideals into more stable and visualized images. The movement from abstract thinking and scattered images to a fully developed blueprint embodying a variety of material and practical considerations follows a process similar to that of the scientist's initial hypothesis to a finally instituted 'fact' that has been corroborated through various scientific operations (Latour, 1987). Just like the scientist's 'theory' embodies the materiality upon which it rests, the architect's blueprint captures – in the ideal case, that is – all available practical concerns and information that need to be taken into account. In summary then, both scientists and architects operate on the basis of the intersection of abstract thinking and concrete and outcomes effects. Their very expertise lies in their ability to apprehend and connect the two.

The importance of talk and communication

In the following, a distinction will be made between on the one hand 'talk', that is, everyday conversations and debates among peers in-house, in the laboratory or in the architecture firm, and 'communication', which is the more formal, more elaborated exchange of ideas with a broader community including both peers in other institutions and firms, but also sponsors, politicians, interest groups and the general public. While 'talk' is the everyday intramural interaction aimed at making sense out of ongoing activities and the puzzles they give rise to, 'communication' is an extramural activity aimed at positioning the focal organization as a legitimate and authoritative actor in a specific field. Again, we draw on the science and technology studies literature. In Lynch's (1985) ethnography of a biology laboratory, the importance of everyday talk is emphasized. Lynch writes:

'[T]alking science' or 'shop talk', is an integral part of the ordinary work of doing science. As such, it is not to be treated as a casual or leisurely activity as, for example, idle chatter during coffee breaks, or as talk which relieves the boredom of routine tasks which can be done 'without thinking'. Although much of the talk that occurs in the lab is such an 'idle' character, 'talking science' is talk which is directly part of the collaborative achievement of inquiry.

(Lynch, 1985: 155).

Lynch also points out that talk is not of necessity only denoting verbal conversations; talk also includes the use of all sorts of resources such as 'written equations, notes, illustrative or analytical diagrams, electron micrographs, oscillographic data inputs, and styrofoam models of lattice structure' (Lynch, 1985: 155). The scientists make use of any means available to make their point. Traweek's (1988) study of high-energy physicists also stresses the need for talk. For Traweek (1988: 117), 'oral communication' is fundamental to the operations because it is what holds the physicists' culture together:

> Physics and its culture is produced and reproduced through talk which is storytelling, talk which is judgmental, talk which punishes and rewards. In short, by gossip ... talk accomplishes diverse tasks for physicists: it creates, defines, and maintains the boundaries of this dispersed but close-knit community; it is a device for establishing, expressing, and manipulating relationships in networks; it determines the fluctuating reputations of physicists, data, detractors and ideas; it articulates and affirms the shared moral code about the proper way to conduct scientific inquiry. Acquiring the capacity to gossip and to gain access to gossip about physicists, data, detractors, and ideas is the final and necessary stage in the training of a high energy physicist. Losing access to that gossip as punishment for violating certain moral codes effectively prevents the physicist from practicing physics.
>
> (Traweek, 1988: 122)

In Traweek's account, talk is not merely aimed at solving problems and making sense out of puzzling research findings, but it is also constituting a physicist culture wherein dominant beliefs and ideologies are expressed and circulated. Several studies of talk in organizations (Boje, 1991; Manning, 1992; Boden, 1994; Donnellon, 1996; Orr, 1996; Czarniawska, 1997; Gabriel, 2000; Kurland and Pelled, 2000; Michelson and Mouly, 2000) show that talk comes in many forms, appears at many occasions, and serves a diverse range of functions; it distributes information, it disciplines and controls, it helps create an *espirit de corps*, etc. Among scientists, talk is what solves problems at hand but also what creates consistency in a field over time. For architects, talk is of great importance because they handle aesthetic objectives and preferences, which of necessity and by definition are complicated to express in exact formulation but demand storytelling skills. In addition, civil-engineering projects and the construction of buildings are regulated in a variety of ways, and numerous agents and groups want to have a say about how the building is being designed. Talk is then a *sine qua non* for making aesthetic, engineering issues and social concerns become fruitfully integrated in the final blueprint. In terms of communication, scientists and architects alike need to persuade various stakeholders that their activities are legitimate and that the outcomes from their respective undertakings could not have been much different given the specific conditions and resources available. As the study will show, the ability to communicate effectively is regarded a major skill of great importance for the performance of an architecture firm.

The importance of external influences

In the common view, scientific laboratories are 'factories' wherein new scientific findings and amazing scientific breakthroughs are fabricated. In reality, scientific breakthroughs of necessity derive from somewhere, but in most cases each laboratory and individual scientists can at best hope to make a rather narrow contribution to a rather delimited domain of expertise (Weber, 1948: 135). That is, laboratories are not only 'factories' but also places where new knowledge produced elsewhere is continuously attended to and integrated into the existing procedures and activities. In Herbert Simon's formulation:

> In any given research laboratory, only a tiny fraction of the new knowledge acquired by the research staff is knowledge created by that laboratory; most of it is knowledge created by researchers elsewhere. We can think of a research scientist as a person who directs one eye at Nature and the other at the literature of his or her field. And in most laboratories, probably all laboratories, much more information comes in through the eye that is scanning the journals than the eye that is looking through the laboratory microscope.
>
> (Simon, 1991: 130)

As a consequence, there is a drive towards standardization across scientific fields; the more standardized a field of research becomes, the more easily new contributions can help in advancing the research front (Fuchs, 1992). Again, Fujimura's (1996) study of the proto-oncogene research provides some useful illustrations. Fujimura argues that the growth of oncogene research can be partially explained by the establishment of ready-made packages of technologies, tools and theories, but also institutional, organizational and political agreements, enabling a quick adaptation of the new research paradigm. Moreover, Fujimura argues that the new 'oncogene package' was successful because it did not challenge previous established paradigms but was instead treated as a complementary approach capable of solving some of the lingering concerns. Fujimura writes:

> [T]he proto-oncogene theory did not challenge the theories to which the researchers had made previous commitments. Indeed, the new research provided them with ways of triangulating evidence using new methods and a new unit of analysis to support earlier ideas. These views of oncogene research were 'realized' through the efforts of these researchers

and, in turn, this realization further extended the reach of oncogene research and the complexity of the theory.

(Fujimura, 1996: 151)

Finally, since no single researcher or research team claimed to have 'invented' the oncogene research, the research was quickly distributed between a number of research centres, thereby giving further momentum to the new field of research. No research project exists in a social vacuum; instead, technologies, tools, methods and theories are circulated between sites and laboratories. Small and incremental rather than major and disruptive changes and contributions are the hallmark of technoscience in contemporary society. Therefore, paying attention to the outside world is a principal virtue of the practicing scientist. The same goes for architect work, at least in architecture firms where creative solutions and new thinking are praised, and where national and international fame and recognition are valued and desired. One of the most common approaches to 'stay updated' with recent work is to read magazines and trade journals wherein the latest contributions are reported and accounted for. Another way to maintain a relation with the broader community is to visit fairs and conferences. Anyway, architects' work is regarded as a form of conversation with others; both with the general public, the architecture tradition and its frontline, and with the actual space wherein the building is constructed. In order to maintain such a conversation, external influences remain pivotal.

Peer recognition and reputation

In many professions and occupations, forms of peer review and peer recognition are of central importance for the identity and performance of the agent (see e.g. Merton, 1973; Latour and Woolgar, 1979; Fuchs, 1992; Owen-Smith, 2001; Nixon, 2005). In the domain of scientific work, various forms of peer review have been instituted since the very birth of proper scientific procedures (Shapin, 1994). In addition, peer recognition is highly valued among researchers. For instance, Hans Selye, prominent stress researcher, points at peer recognition as the gold standard of any scientific field:

Many of the really talented scientists are not at all money-minded; nor do they condone greed for wealth either in themselves or in others. On the other hand, all the scientists I know sufficiently well to judge (and I include myself in this group) are extremely anxious to have their work recognized and approved by others. Is it not below the dignity of an objective scientific mind to permit such a distortion of his true motives? Besides, what is there to be ashamed of?

(Hans Selye, cited in Merton, 1973: 400)

The importance of peer recognition may be explained by the different functions it plays. First, most scientific work is more or less abstract and its outcomes and practical consequences are often obscure or complicated to foresee. Peer recognition is here a means for reducing uncertainty and the anxiety for being 'off the track' such uncertainty would entail. Second, peer recognition may also be positively related to the ability to attract funding and to achieve one's earned reputation. Fuchs (1992: 72) speaks of reputation (i.e. peer recognition) as what is helping scientists to navigate in territories wherein many statements are being made: 'Statements made by scientists with high reputation have better visibility and hence are more likely to be recognized by other scientists ... No scientists listen to all statements. Reputation reduces complexity for scientists whose limited span necessitates decisions about whom to listen to, whom to ignore, whom to ridicule, and whom to take very seriously'. Therefore, being recognized by peers as an important maker of 'statements' (i.e. formulated scientific facts) is what helps the scientist stay competitive in a particular field of research.

In architectural firms, peer recognition plays a similar role, both internally and externally. Internally, co-workers have to prove themselves worthy of managing a project or being capable of conceiving of creative solutions to identified problems. When passing the tests of senior architects and managers, more interesting and challenging work assignments may be the reward. Externally, individual architecture firms and especially individual architects may be recognized on the basis of their historical track records. Ivory points at the value of getting one's architecture reviewed and recognized:

> Clearly, for the architect, reputations are tied to previous works. Every building is, as one architect noted, 'a sort of database, a living advertisement'. In other words, these buildings as they appear in portfolios and in presentations, become key to winning future business. Innovation becomes central to the process because a building which looks good, even if it is only because of an innovative roof or cladding system, will always speak better of the architect than a more mundane creation.
>
> (Ivory, 2004: 506)

Individual famous or widely recognized buildings may be the sign of excellence of an individual architect throughout his or her life. The perhaps best known example is the Danish architect Jørn Utzon's Opera House in Sydney, not only a symbol for Utzon's own work and competence, but for Sydney itself and, at times, even for the whole of Australia. The field of architecture – just like science – is a field of a few superstars, a number of local celebrities and many only marginally recognized laboratory/office workers. Just like the sciences have Copernicus, Newton and Einstein, architecture has Paladio, Le Corbusier and Frank O. Gehry. The downside of the peer-recognition culture is the resentment felt when individual contributions or collaborations are not sufficiently recognized. For instance, in Rabinow's (1996) study of the biotechnology firm Cetus, which developed the polymerase chain reaction (PCR) – today an essential technique in biotechnology research – several co-workers felt left out when Kary B. Mullis received the Nobel Prize in Chemistry in 1993 for this contribution. One of Mullis's disgruntled collaborators deplored the inability to recognize joint work: 'Committees and science journalists like the idea of associating a unique idea with a unique person, the lone genius. PCR is, in fact, one of the classic examples of teamwork' (Henry Erlich, Cetus scientist, cited in Rabinow, 1996: 161).

In summary, scientific work and architect work is a form of materialized semiosis, operating on the border between the abstract and the concrete, the theoretical and the material; it is based on the ability to talk and communicate internally to the laboratory or architecture firm as well as with external stakeholders; it is based on the ability to absorb information and work conducted outside the focal site or office; it is embedded in a peer-recognition culture wherein it is the review and recognition of professional peers that is the true gold standard rather than the outsider's view. It is nice to be recognized by outsiders but it is what peers believe that truly matters.

Architectural work: from sketching to materiality

The organization and leadership of the architects' office

Blue Architect Firm is an architectural consultancy firm founded by the chief architect and chief executive officer (CEO) Lars Andersson in the second half of the 1970s in Gothenburg, Sweden. The firm has grown organically and today employs about 100 at its offices in Gothenburg and Stockholm, Sweden. The company became known internationally after being represented at the architecture exhibition at the 1996 Venice Biennale and after winning the competition for the Swedish embassy in a major European capital. Today, Blue Architect Firm is one of the most respected in northern Europe and Lars Andersson has been awarded a number of prestigious architecture prizes. As a consequence, Lars Andersson is, if not a celebrity, well-known among the general public interested in architecture in Sweden. Recently, the company has won several international prizes for, amongst others, a number of Swedish embassies, a corporate headquarters in London, and a national science centre in Gothenburg. Blue Architect Firm has also won a 'Laboratory of the Year' award in the United States.

For his co-workers, Lars Andersson is a great visionary and leads the day-to-day activities at the office without being involved in all the details. Andersson claimed that he could 'allow himself to act a bit like an elephant in the office' even though that may 'repel some talents' because it is the mutual interest in producing good and interesting architecture that matters at the end of the day. At a single point in time, there could be as many as 40–50 concurrent projects running at a time, and therefore Andersson is not involved in all the details. Instead, he participates in the more creative phases where concepts are developed and where new thinking is articulated. In the more highbrow projects, such as the embassy buildings, Andersson plays a more active role. Failing to produce qualitative solutions in such projects would be detrimental for the architecture firm, and thus Andersson invests more time and energy.

The project work procedure includes at least three distinct phases. First, the *competition phase*, wherein the architecture firm is preparing its contribution for competition on basis of the specifications and instruction provided by the client organization. Second, if the architecture firm wins the competition, a *system-specifications phase* takes place where all sorts of documents are collected, and legal documents, formal decisions and contracting are completed. Third, the *production phase*, wherein the actual building is constructed. The third phase includes dealing with emerging problems occurring as the building is constructed. Throughout the process, the architects collaborate with a great number of professional and occupational groups (Ivory, 2004). Architectural work is thus a mixture of science, civil engineering, visualization and art bringing together a variety of competencies. This work also includes a series of practices. The architect *reads, sketches, visualizes, talks* (internally), *communicates* (externally); not only is the form of the (emergent) building attended to, also its matter, its materiality, needs to be controlled and designed by the architect. In addition, the architect is not responsible for the very construction of the building, but he or she merely provides the blueprints and other specifications used in the production phase. This entanglement between materiality and symbolic resources, sharing many characteristics with scientific work, is examined in the following section.

Blending the symbolic and the material

Similar to laboratory scientists, architects operate on the boundary between the symbolic and the material. Their creative solutions to articulated problems are represented by sketches, blueprints and computer-generated images that are on their way to materialization in the production phase of the construction project. In addition, the expertise of architects lies in their ability to translate generally formulated instructions and specifications into 'workable' solutions that optimize a variety of objectives, at times competing with one another. The chief architect explained the work procedure and pointed at his own role in the process:

> There is also a rather substantial body of research to draw on. That information is turned into a graphic scheme and that graphic scheme is what one may call a site plan or a section structure of the buildings.

I am not doing that work or somebody else highly skilled, but somebody very careful who won't miss any details. My competence is the graphical space design. I think that is the core of the architect's profession.

(Chief architect)

The process wherein written texts and verbally articulated practical concerns are translated into graphic form is referred to as *sketching*. 'We talk about the *sketching* – You "sketch a project". That is the creative process when you work in the project, the sketching phases', one of the architects said. One of the senior project leaders pointed out that a variety of competing objectives had to be brought together in the sketching process:

You sketch all the time. You sit down and seek to figure out what is important and, as always, not everything is possible to combine. All demands and wishes cannot be unified and you cannot tick the boxes like 'we've handled this and we've handled that'. It is all about making priorities. These things can be solved and those solutions have these consequences. It is a form of tinkering throughout.

(Senior project leader)

One of the architects emphasized that the sketching phase is at times frustrating when there is little time or no useful ideas emerge:

It can be incredibly frustrating to work in the sketching phase; you think 'Shit! we have no good ideas'. It is a real drag. But then all of a sudden you think 'Wow, this is good! This idea can be carried through the entire project'. Then it is just great. The one day is pure torture while the other is amazing.

(Architect)

He continued: 'In the worst case scenario, there is little time and then you may re-use old ideas from previous projects or so on. That easily happens. "That solution for the stairs worked fine, we'll use that again". Such things happen, right'. An important component of the architecture ideology is the idea that architecture is not solely a matter of the individual building but the ability to locate a specific building in an individual space. What 'makes a difference' for the architect is then the ability to explore what is called the *genus locus*, the idiosyncrasies of the particular space:

It [the building] is supposed to be different. It needs to be part of the environment, yet it needs to be something new. In architecture we speak of genus locus, the very 'soul of the place'. If you take a pre-fabricated building and locate it to a certain place then you notice that this house is not designed to suit this place. That is really important.

(Architect)

It is the ability to combine the symbolic and the material of the *locus* in a creative manner that is the mark of the skilful architect. Lars Andersson pointed to the ability to conceive to the unexpected as a principal competence:

> We have an internal motto – our mission is to give the client what he didn't know he wanted to have. This is to indicate that something beyond the standard solution is needed for a certain task. To be honest, I would say that architecture is an innovation profession not a creative profession, because we are constantly manipulating known components and elements. Being creative would be to take away the pillars making the house float in the air – we don't do that very often, especially not with good results. Architects' work is a craft with something extra. Architecture, as a domain, quite rarely develops new materials. Architects in general are not very aware of different construction materials, or deep domain knowledge in technology aspects.
>
> (Chief architect)

This aspect is well illustrated during the preparation and design of the Swedish National Science Centre opening in Gothenburg in 2001. The actual building is more than 10 000 square metres in area and contains aquariums, experiment stations, exhibitions and information technology-based education solutions. It is notable that wood, glass and concrete are the three main materials employed. If at some time in the future the building has to be knocked down, this is made easy by the construction solution and all the building materials can be recycled. The National Science Centre has been built to be an ecological role model in every way. The architect responsible for the project explains:

> In the case of [the National Science Centre], the first idea was a collage of different parts and a connection with different landscapes – the rainforest was to be connected with the actual exterior through a large set of glass windows. The key point was that the perception of the rainforest should be larger than it actually was when contrasted with the external surroundings of the building. This idea came early because the surroundings were good. Then different parts emerged but details hadn't yet found their right shape ... I presented the second concept to Lars and he thought it may work. He then worked it over and faxed it back to me combined with the new concept of using large pieces of wood in the building. This process included a number of iterations back and forth. The idea of the pronounced large wood entrance derived partly from the environmental aspect, which was already described in the competition requirements. Our first intention was to build the large roof by using an ordinary steel construction. But in the continuing sketch process, the concept of using wood not only for the entrance

but also for other parts emerged. Then we decided that a new point of departure would be a building with very general and simple design with great flexibility. And this fitted into the wood concept. We wanted to show that it was possible to build a modern building in wood, instead of concrete, something that is actually quite rare nowadays. This wood construction was actually much more expensive than steel, but it represented more originality, and was in line with the ecological objectives. Although more expensive at the time of construction, from a life-cycle perspective, it will become cheaper.

(Architect)

One of the architects emphasized that the creativity in the work lay in the ability to make decisions on the basis of different and, in many cases, incomplete information:

Creativity is like thought-processes, ways of making decisions. It may be very different from one time to another; you use different sources of inspiration or you have different ideas on how to design the house. There are different types of thought-processes and it is complicated to capture them; they differ substantially from project to project.

(Architect)

In many cases, architect work is not solely about conceiving of aesthetic solutions but also about handling very practical and mundane matters. 'Aesthetic matters can be discussed endlessly: What looks good and what doesn't. But one needs to agree this is the right approach', one of the senior project leaders suggested. Architect work is then a form of tinkering or bricolage wherein heterogeneous objectives and concerns are brought together and aligned. The line of demarcation between the symbolic (e.g. written documents, graphs, sketches, computer-generated images, etc.) and the material (e.g. the location of the forthcoming building, the choice of materials) is continuously transgressed in architects' work.

Talk and communication

Another feature of architects' work is that it is to some extent a joint accomplishment embedded in the ability to what we have referred to as (internal) talk and (external) communication. Among the interviewees, these two forms of communication were regarded as being of central importance for the activities. One of the computer-visualization experts argued that creativity was inextricably entangled with dialogue and conversation:

Creativity is in the dialogue, in the conversations. There is a communication between image and building. The architects are the ones designing the building and I think I am the most creative when I discuss with the architect in the project. Clearly so. To articulate what I think and so forth. I sketch nothing by hand. The computer is my tool and you may say that you are creative when you sketch. But where I really think I am creative, that is when I discuss with others.

(3D-visualizer)

One of the architects emphasized the ability to share ideas on architecture as one of the most important leadership skills in the domain of architects' work:

Besides the creative abilities, leadership skill is to be able to communicate this idea further, to explain to others how you thought and engage the co-workers to think in similar terms. A good leader is capable of making the co-workers believe this is their own idea, one may say, and then you can work towards the same goals. That is very important; it is pivotal for the creative processes in the office.

(Architect)

In the Gothenburg office, the design of the office space promoted a continuous interaction between project co-workers. The chief architect stressed that an open flow of information and know-how helped reduce time in the project:

We know from studies of Ericsson [the telecommunications company] and other places, that what prevents work from proceeding is lack of information, not the lack of separated workspaces. Sitting in close proximity and overhearing one another and meetings, that is a way of sharing information that is good for the projects.

(Chief architect)

Some of the interviewees pointed at the growth of the firm as an obstacle for such free-floating information. The computer-visualization experts reflected on the changes over time:

I have been here for seven years now so I've been here since we were not even 20 people, maybe 15, when I started. Of course, that is a real difference. For instance, you don't know everyone the same way. We are located on four different floors. And the organization is gradually becoming slower, to get things done, changes in the corporate structure, and so forth, that is a source of complaints. Inertia becomes an issue.

(Chief architect)

Besides sheer space and increased distance between individual co-workers, the chief architect saw the ability to communicate with the help of computer-generated images as one of the bottlenecks for the company:

> The single most significant challenge for developing the company today
> ... is that we spend too much time communicating. ... We have four
> dedicated data graphic visualizers and we need even more because we
> communicate with images supported by texts. Today, such skills are the
> bottleneck. That is the group working most overtime today.
>
> (Chief architect)

Internally, talk served to constitute a joint vision in the project, to motivate co-workers, and to create a sense of *esprit de corps*; externally, communication played a similar role, albeit with other stakeholders not of necessity sharing the architect credo. Several co-workers praise the chief architect Lars' ability to communicate effectively with stakeholders. One of the data-visualization experts said: 'I think Lars' personality plays a substantial role. His way of communicating and convincing people and making them go for something that makes a difference matters, I would say'. One of the design engineers argued in the same vein:

> Lars is great at promoting his ideas. He is very eloquent. I think that
> is number one. Moreover, he does not recognize any limitations. If he
> wants it a particular way, then he works hard to accomplish it and
> makes the co-workers join him in the work.
>
> (Design engineer)

Similar to all entrepreneurs, the chief architect Lars had developed skills to persuade various stakeholders to follow his visions. Communication is then not only a matter of sharing information but also a means for joint storytelling wherein possible futures are formulated and elaborated upon.

The importance of external influences

In analogy with the scientific laboratory, the architectural firm is a site paying minute attention to new ideas and new influences from the outside. The laboratory and the architects' office are thus permeable and porous social organizations demonstrating a fundamental openness to the outside. Just like scientists may 'jump the bandwagon' (Fujimura, 1996), architects may draw on and exploit ideas formulated elsewhere. One of the architects argued that there were in fact a number of architect firms they saw as role models: 'There are certain architects that are known for being innovative. It may be that we are recognized too, but we are very influenced by others – we have many role models for our work' (Architect). Also the chief architect, himself a recognized and widely known contributor to contemporary architecture, relied extensively on external sources of inspiration. One of the co-workers argued:

> A lot of inspiration comes from others. He [Lars, the CEO and chief architect] read incredible amounts of magazines. He knows what kind of feelings he wants a building to convey. If he cannot explain in words what he wants to say, then he goes to his room to pick up a magazine and then we have an image of what he wants it to look like. You know, what type of stone to use on the façade and so forth. His creativity is no doubt an inspiration. I can imagine that most people being creative in design professions are inspired by many sources.
>
> (3D-visualizer)

When being asked in detail about his source of influence, the architect drew on a vocabulary, which, out of necessity, outsiders can relate to, for instance, the 'architectural concept':

> Q: So what you're saying is, it's not only the final building that matters?
> A: No, it is about the entire architectural concept; how you approach an assignment, how you find what is the essence of the assignment and creates something that has not been conceived of previously. But novelty as such is never a goal in itself. ... You must not invent something for its own sake. What I mean is, one needs to find the very point with the assignment.
>
> (Architect)

Lars, the chief architect, thought of the various architecture journals as the principal means of communication in the field: 'There is a continuous publication of model photos and blueprints in architecture journals; that is our communication with one another', he argued. The journals thus played a role similar to the double-blind review journals used to report scientific findings in scientific communities. Failing to participate in the conversation in the global architect community means to miss an opportunity to stay updated. Again, architect work is not a matter of isolated activities, but of a systematic engagement with problems and challenges that are both local and global, both symbolic and (becoming) material.

Peer recognition

Since architects' offices benefit from a good reputation and can 'transform' credibility into new assignments and financial performance – for instance, certain competitions only invited contestants, and credibility was the principal criterion when selecting architecture firms – the whole field is, similar to scientific work, preoccupied with peer recognition. While researchers publish research findings in peer-reviewed journals of different ranking and influence, architects gain credibility through winning architecture competitions and when their work is being published in architecture magazines. Peer recognition and recognition in general (what may be called reputation in the broader society) are then the 'gold standards' of the industry (Rabinow, 1996). When asked about motivation, the interviewees tended to make references to forms of recognition. The 3D-visualization expert said:

> I would say, doing things that makes a difference. I don't know what to call it, to create something new. I think that is a major source of motivation for most of us, to contribute with something that is recognized and reviewed and that is credible.
>
> (Chief architect)

One of the senior project leaders claimed that the chief architect's visions were influenced by ideas of how to achieve recognition among his peers and the wider public: 'It is his [Lars Andersson's] visions that one needs to pay attention to. To produce the largest and most beautiful buildings in Sweden and abroad. To get famous. Become recognized'. Andersson himself was very satisfied with the accomplishments of the office to date:

> We used to say [ten years ago] that we would become a leading Swedish architecture office, inspired by the best international role models. That was an unusually clearly formulated vision and we may say that we have accomplished this vision. After that, it has been much more complicated to formulate a new vision because it is more complicated to communicate a new one. Now we say we want to become the leading architecture office of the Nordic countries and that we have not yet accomplished.
>
> (Chief architect)

Lars Andersson also emphasized that winning prestigious architecture competitions was the key to recognition in society and in the international architects' community. The two prestigious Swedish Embassy buildings were, for instance, eminent examples of how to gain recognition in the international architectural press. Lars Andersson argued:

> The entire community of architects are continuously in contact with the architecture journals and there is an institutionalized market ... There is little doubt that if you do a project in a 'visible space', then you get more attention, right. Our most widely published project, that is the Swedish Embassy in Berlin because Berlin as a city of architecture is closely attended to. All of the architecture press covers it. It is published much more than, for instance, the Museum of Modern Art in Stockholm – not because it is better – but because of the location.
>
> (Chief architect)

Being in the right spot is then more important than producing a spectacular building in the periphery. Similar to the science community and its strong emphasis on ranking publications and journals and tracking citation patterns, playing a role internationally makes a difference between international recognition and obscurity.

In addition to the external recognition, in many cases sticking to the chief architect being primarily the formal and official representative of the office, there is an intramural market for credibility that needs to be attended to. One of the senior project leaders argued the importance of internal competition: 'It is a young team, many of them want to accomplish very much and I think people push one another to become more creative. It is a kind of a competition between some co-workers'. The chief architect himself was very aware of the need for acknowledging individual contributions to the project, especially those attracting much attention in the broader community: 'Not getting any recognition if you have contributed to something we have published, that is serious stuff', he said. As a rather integrated and fairly stable profession, architects are dedicated to the task of designing buildings, but their day-to-day work and performance are also dependent on their ability to gain recognition. Like most creative professions, a combination of intrinsic and external motivation is the fuel propelling forward equally individual careers and the industry.

Architects' work as a 'mangle of practice'

Owing to recent interest in the use of know-how, skills, competencies, expertise and other forms of knowledge in organizations, practices not previously examined in detail are now being explored. For instance, the field of scientific work has been approached from an ethnomethodological perspective or in terms of being an activity based on its ability to mobilize actor networks. In such studies, scientific work and, more specifically, laboratory work is envisaged as an ordering and structuring of an experimental system comprising technologies, tools, theories, practices, organic materials or laboratory animals, such as genetically modified mice, and so forth, in order to 'prove' or 'stabilize' a theory or a theoretical framework (Pickering, 1995; Fujimura, 1996; Rheinberger, 1997). Scientific work is then a series of complex social interactions wherein a variety of heterogeneous resources are enrolled and take part. One of the principal characteristics of contemporary technoscience (i.e. science and technology constituting an assemblage not very meaningful to separate from either an analytical or a practical perspective) is that it produces a particular form of 'nature' that is, at best, loosely coupled with the 'naturally occurring' nature outside the laboratory domain. Scientific work thus operates in the borderland between the theoretical and symbolic and the material and practical. In addition, scientific work is anchored in a range of instituted practices such as the peer-review routine and the (supposedly) free circulation of research findings.

In this chapter, architect work has been examined on the basis of the extensive corpus of research on scientific practice. Architects' work follows, it has been argued, some of the trajectories demonstrated by scientific communities. Architects' work is what bridges the symbolic and the material, what is dependent on internal talk and external communication, is drawing on accomplishments and performances produced elsewhere, and is fuelled by various forms of peer recognition. There are, however, some notable differences that are worthy of being addressed. First, scientific procedures are generally standardized and thus allow for comparisons between universities, nations and continents. The academic emphasis on the ranking, evaluating and reviewing of one another's work has produced a situation wherein one can say, with fairly high predictability, who are the leading scientists and which are the leading laboratories within a specific field at a particular point of time. In architects' work, there are no such universal and generally agreed standards for evaluations. Local traditions affect architects' work. Of course, there are respected architects, famous buildings, reputed architects' offices and so forth, but the control and evaluation is not as tightly coupled as in the field of the laboratory sciences or natural sciences (Fuchs, 1992). Second, architecture is at the bottom line, dependent on what Fine (1996), studying restaurant chefs' work, speaks of in terms of aesthetics (for a detailed discussion, see Chapter 5). Fine explicated this most complex concept:

> This concept [aesthetics] is the broadest of a cluster of terms that involve sensory qualities of experience and objects; beauty, creativity, elegance, goodness, and the like. An aesthetic object, or act, is intended to produce a sensory response in an audience ... No special brief exists for this definition other than its utility and general reasonableness. It captures the cognitive ('satisfaction') and affective ('sensory') components of aesthetic judgments and includes the intentional quality of human action. Aesthetics emphasizes that these choices are distinct from purely instrumental and efficient choices: the workers care about 'style', not only about technological quality. Although form and function are typically intertwined, aesthetics refers specifically to the production of form, not only function.
>
> (Fine, 1996: 178)

For Fine, the restaurant chefs' work is entangled with their aesthetic choices and preferences. Aesthetics is for Fine (1996) an 'activity' or a 'social accomplishment' based on such cognitive and sensory skills, more or less tacit. Of course, there are traces of aesthetics concerns in scientific practice but in general researchers are too preoccupied with making their experimental systems work properly to have time to bother about aesthetics. Architects are, on the other hand, trading their ability to review, evaluate and contribute to a social practice, recognizing the complex category of aesthetics.

Taken together, there are, however, more similarities than differences between architectural work and scientific practice. First and foremost, both architectural work and scientific thinking are based on the ability to use imagination on how to provide a solution to a perceived problem. Koyré (1968: 45) speaks of the influence of what Ernst Mach called *Gedankenexperimente,* 'thought experiments' in scientific thinking. Thought experiments are not real experiments, including technical apparatus and other scientific resources, but are instead thought trails wherein possible explanations for a puzzling phenomenon are examined. Similar to scientific work, architects' work draws on the architects' abilities to engage in such 'what-if thinking', conceiving a variety of solutions to identified and perceived problems. Moreover, both scientists and architects are eminent 'cartographers', capturing their thought experiments by various graphs, models, illustrations, models, equations and so forth. Again, the two domains of expertise are mixing abstract thinking, various practices for graphically illustrating such thinking and anticipating the material effects of this work. Laboratory researchers formulate hypotheses and test them in their experimental systems; architects discuss solutions to problems and sketch a number of solutions and have them visualized as computer-generated images. Neither of these practices are yet materialized but they are on the way to taking a more immutable material form. Scientific findings are generally claimed to have practical effects and are often announced as biopolitical statements the general public is expected to pay attention to; the architects' work is eventually becoming a physical building made out of concrete, wood, glass panes, various technological systems (e.g. elevators, sprinkler systems, and surveillance technologies), and other material resources. The point here is that both scientists and architects inscribe the world in documents that on the basis of the very credibility and expertise of the two groups will entail material consequences. Both scientists and architects engage in what we here, following Haraway (2000), call materialized semiosis.

In the general interest for knowledge management and how organizational capabilities are developed, the tradition of science and technology studies and its emphasis on detailed accounts of work practice and the seemingly mundane and peripheral aspects of everyday work-life may serve as a fruitful influence. Rather than assuming that a particular profession or occupation engages in clearly demarcated activities and in a linear manner, the day-to-day 'mangle of practice' (Pickering, 1995) and its putting together of operating systems of practice deserves to be examined. Knowledge management and other forms of research programmes that aim at understanding the way know-how is used in its local and situated context benefit from detailed and sensible accounts of practice(s) in various fields. Rather than addressing professions or occupations *en bloc*, every such domain explodes into a thousand tiny practices glued together by more or less explicit ideologies, shared beliefs, collective identities, and hopes for the future.

Summary and conclusion

This chapter has reported a study of an internationally renowned architecture office and suggests that architects' work shares a basic morphology with scientific work: architects' work is conducted in the intersection between the symbolic and the material; it is based on the ability to talk and communicate both within and outside the firm; the main influence for new thinking is derived from outside the focal site; and, finally, architectural work is propelled by the distribution of peer recognition and credibility on the basis of individual and collective performances. However, this does not suggest that architectural work should mimic scientific work or that these particular characteristics demonstrate some intrinsic rationality. Instead, these are some of the traits of two fields wherein ambiguities, fuzzy boundaries, and performances that are not always easily measurable need to be mediated by certain institutions and ideologies. What is of particular interest in this perspective is that architectural work is a social practice based on individual and collective contributions and the shared work to make sense out of a variety of activities. Therefore, one must not conceive of architectural work as a black box evolving as a series of ready-made and essentially unproblematic activities. Instead, architectural work is, like any complex social and knowledge-based work, an ongoing attempt at aligning a variety of heterogeneous resources and practices.

4 The architect's gaze

Visual artefacts, perception and knowledge in architects' work

Introduction

In the literature addressing how various skills, know-how and competencies are mobilized and put into use in organizations, there is surprisingly little emphasis on the use of the five senses, that is, perception in knowledge-intensive work. Knowledge is often assumed to be cognitive and located in the cerebral structure of the brain, and based on a series of propositions, statements of factual conditions of the world that may be applied to cases (Boisot, 1998; Teece, 2000; Bontis, Crossan and Hulland, 2002). In an alternative perspective, knowledge is distributed, fluid and socially embedded (Gherardi, 2001; Tsoukas and Vladimirou, 2001; Tsoukas, 2005; Chia and Holt, 2008) and is the outcome from the combination of a number of human faculties, including the ability to see (Belova, 2006; Edenius and Yakhlef, 2007), listen (Porcello, 2004), touch (Prentice, 2005), taste (Fine, 1996) and smell (Fitzgerald and Ellen, 1999) entities in the social reality and cognitive capacities. In this view, knowledge is not what exclusively resides in cognitive processes but is instead what emerges when a variety of human skills are integrated and combined. For instance, Mody (2005) suggests that scientific procedures in laboratory work are based on the researchers' and laboratory technicians' ability to listen to and interpret the sounds of the laboratory equipment to determine whether the technology is operating as intended: [S]ound is an integral (if often over-looked) ingredient in tacit knowledge. Surface scientists carefully manage auditory (as well as visual and haptic) cues to liberate different kinds of information from the experiments', Mody (2005: 177) suggests. In addition, more specialized professional groups, such as recording-studio technicians and sound engineers use their capacity for hearing and evaluating sounds produced within a an assemblage of technologies as an integral component of their everyday work (Porcello, 2004).

While auditory capacities are useful in many cases in a variety of occupations and professions, the faculty of vision is perhaps the single most important perceptual capacity in organizations. A number of social theorists and organization researchers have examined the concepts of vision and visuality in society and in organizations. While all uses of perceptual capacities are to some extent trained and developed in professional settings – for instance, becoming a *sommelier*, a wine expert hired by some upscale restaurants demands significant training and experience from tasting, smelling, discussing, and evaluating wine – the capacity of vision is in many cases taken for granted and rendered unproblematic. However, as has been suggested by a variety of writers, the capacity to observe and inspect qua professional or representative of a particular occupational group is a skill that is acquired through years of training. Medical doctors are trained to diagnose symptoms; police officers learn to evaluate certain human behaviours; laboratory researchers develop the skill of inspecting images of graphs being produced by the technical equipment in use (Law and Whitttaker, 1988). All these different forms of visual inspection are referred to as 'professional vision' by Goodwin (1994, 1995, 2001). Professional vision is part of the skilled professional's repertoire of practices and plays a central role in a large number of knowledge-intensive organizations. Professional vision is a form of vision already anchored in professional norms, values and beliefs, and the professional onlooker is therefore already part of a visual community. Following Jacques Lacan, this visual community is subject to a particular gaze, a 'way of seeing' that is inherently structured by instituted professional beliefs and assumptions. This chapter examines how the visual community of architects draws on what may be referred to as the architect's gaze, that is, the collectively accomplished skills of visually inspecting a social space or surface in a standardized manner. Just like the sound engineers studied by Porcello (2004) shared a set of onomatopoetic cues, standard references, and clichés to denote particular sounds when discussing recorded music, the community of architects draws on a shared vocabulary when examining social space and surfaces. The architect's gaze is therefore what is central to the architects' profession in terms of offering a repertoire of visual practices and vocabularies that may be drawn on when discussing space and surfaces with both colleagues and outsiders.

Perception and knowledge: the gaze, professional vision and visual artefacts

Studies of architects' work in organization theory often examine the work procedures and the professional traits of the architects (Blau and McKinley, 1979; Blau, 1984; Gutman, 1988; Cuff, 1991; Winch and Schneider, 1993; Pinnington and Morris, 2002; Andreu and Oreszczyn, 2004; Yaneva, 2005; Boland, Lyytinen and Yoo, 2007). In a few cases, procedures pertaining to vision and visuality and, more specifically, the use of visual aids in the work (Ewenstein and Whyte, 2007a, 2007b; Whyte, Ewenstein, Hales and Tidd, 2007) are addressed. For instance, Bechky (2003b) reports how engineers and adjacent professional and occupational groups are negotiating their domain of jurisdiction by using drawings as 'epistemic objects' in their joint boundary-work. Henderson (1999) and Bucciarelli (1994) study how engineers are trained to use visual representations such as sketches as an integral resource in their thinking and engagement with practical problems. 'I can't think without my drawing board', one engineer exclaimed (cited by Henderson, 1999: 1), testifying to the proximity in engineering between sketching and thinking.

Architects are often portrayed as a professional group standing midway between, on the one hand, the material and overtly practical world of the construction industry and the built environment and, on the other hand, the aesthetic, visual and in general more abstract concerns of the world of designers, art critics, social planners and so forth. Architects often recognize the aesthetic components in their work but at the same time take pride in making a contribution to the immutable social order manifested in the built environment. One important skill in bridging the world of concrete materiality and abstract aesthetics is the ability to inspect and examine sites, buildings and architectural solutions to perceived problems visually. That is, the concept of vision and visuality is of central importance for the architectural profession.

Jonathan Crary (1990, 1995, 1999) suggests that it was not until the first decades of the nineteenth century that a 'modern' theory of vision was established. In the classic regime of visuality, visual perception was essentially accomplished in the same manner for all human beings. In the emerging new perspective, vision was instead treated as a subjective skill and is therefore what may be trained:

> The idea of subjective vision – the notion that the quality of our sensations depends less on the nature of the stimulus and more on the makeup and functioning of our sensory apparatus – was one of the conditions for the historical emergence of notions of autonomous vision, that is, for severing (or liberation) of perceptual experience from a necessary and determinate relation to an exterior world. Equally important, the rapid accumulation of knowledge about the workings of a fully embodied observer made vision open to procedures of normalization, of quantification, of discipline. Once the empirical truth of vision was determined to lie in the body, the senses – and vision in particular – were able to be annexed and controlled by external techniques of manipulation and stimulation.
>
> (Crary, 1995: 46–7)

In the new theories of vision, the body becomes a sensory-motor system that in various ways shapes vision. As a consequence, concepts such as *attention*, apparently differing between individuals and situations, became subject to laboratory research by, for instance, Hermann von

Helmholtz in the mid-nineteenth century (Crary, 1995: 51). The entire aggregate of vision, visuality and attention became a veritable domain of scientific research from the mid-nineteenth century onward.

In Jacques Lacan's psychoanalytical theory, influenced by Merleau-Ponty's phenomenological philosophy, vision is not only subjective but is instead what is always already shaped and formed by 'the gaze of the other'. Lacan (1998: 78) speaks about the 'split between the gaze and the eye'. There are no autonomous and primary positions of the viewer enabling 'pure vision' but vision is from the outset influenced by power and becomes visuality under the influence of dominant social forces and interest. Merleau-Ponty (1968: 83) expresses this idea in terms of vision having its specific limitations: '[t]he privilege of vision is not to open *ex nihilo* upon a pure being *ad infinitum*; the vision too has a field, a range'. Visual perception is then by no means original or detached from broader social interests but is instead what is inscribed with meaning and function. In Bryson's (1988) formulation:

> Lacan's analysis of vision unfolds in the same terms: the viewing subject does not stand at the center of a perceptual horizon, and cannot command the chains of signifiers passing across the visual domain. Vision unfolds to the side of, in tangent to, the field of the other. And to that form of seeing Lacan gives a name: seeing on the field of the other, seeing under the Gaze.
>
> (Bryson, 1988: 94)

The perceiving subject is already submitting to what may be seen within a particular community and in a particular regime of visuality. The gaze is thus the vision that is of necessity shaped and influence by predominant beliefs and assumptions. Speaking of professional vision, the gaze does not of necessity mean that vision is already corrupt and deceiving. Instead, Lacan suggests that the gaze is what must be understood in psychoanalytical terms and what is in essence social in nature. The gaze is therefore what is, among other things, developed within professional communities and in professional training and work.

The concept of gaze has been used in various domains of the social sciences. It was first used by Michel Foucault (1973) in his study of the emergence of a medical profession, characterized by a certain 'medical gaze' [*regard médicale*] enabling the diagnosis of the patient. For Foucault, the medical gaze includes all of the skills, know-how, norms and values of the experienced medical doctor. The concept of gaze has also been very frequently discussed in media and cinema studies since Laura Mulvey (1989) in the mid-1970s spoke of the 'masculine gaze' turning women into objects. 'In a world ordered by sexual imbalance', Mulvey (1989: 19) writes, 'pleasure in looking has been split between male/active and passive/female. The determining male gaze projects its fantasy onto the female figure, which is styled accordingly'. More recently, sociologists like Norman

Denzin (1995) have used the concept of the gaze in a general critique of the contemporary society as what is 'cinematic' and characterized by the ubiquitous presence of the 'voyeur's gaze' (see also Beller, 2006). Entwhistle and Racamora's (2006) study of the London Fashion Week is an example of such cinematic '*mise-en-scène*', drawing on a particular 'fashion gaze', where both the models on the catwalk and spectators are operating under the influence of the gaze:

> The catwalk theatre is a particularly visible realm where identities are created through the very visible performances, which in turn constitute part of the way in which struggles in the field are played out. The staging of the catwalk shows a staging of the gaze: the gaze of the participants sitting in the audience, who are once its objects and subject. This gaze contrasts with that of the models, distant and detached, a gaze that does not watch, that is there not to see but only to be seen.
>
> (Entwhistle and Racamora, 2006: 744)

'Whoever fights monsters should see to it that in the process he does not become a monster. And if you gaze long enough into an abyss, the abyss will gaze back into you', Nietzsche writes in *Beyond Good and Evil*. This is how the gaze operates: the look of the spectator is at the same time turning back on its subject. Both the object and the subject of the gaze are included in the specific regime of perception instituted by the gaze (Denzin, 1995: 44). Using Mulvey's (1989) example of the cinematic gaze, the male spectator is not only observing women as sexualized objects but is also looking into himself, into the subject that is ultimately a product of a patriarchal regime which neither he nor the female subject can escape. The gaze is the instituted modes of perception or ways of seeing in a particular social formation. While, for instance, Michel Foucault has insisted on the disciplinary function of the gaze (especially in his so-called 'middle period' of the panopticon writings, in for example, *Discipline and Punish*, 1977), the power inherent to seeing and inspecting, Žižek (1995) has on the contrary emphasized the powerlessness of the gaze, the predicament of observing without being able to intervene or take proper action, effectively illustrated by Alfred Hitchcock's *Rear Window*, where the character played by James Stewart, stuck in his wheelchair, is observing a murder in the neighbouring house. The gaze is what is both operating as a means of power, but can also be a source of despair and sense of helplessness.

The gaze of professional communities may be referred to fruitfully with the term 'professional vision' suggested by Goodwin (1994, 1995, 2001). For Goodwin, the professional vision is composed of a series of activities. First, professional vision is helping the observer *codify* what is inspected. Second, specific features of what is subject to inspection are *highlighted*, that is, important aspects of the inspected object are sorted out as being of particular importance. Third, *material representations* are produced or articulated on the basis of the codification and highlighting (Goodwin, 1994: 606). As a consequence, the members of a professional community are capable of shaping the way an object of inspection is examined and discussed. Professional vision is therefore what is both a 'way of seeing', a specific 'gaze', but also an accompanying vocabulary on how to speak of, relate to, discuss, address and otherwise articulate what has been observed. For instance, in Goodwin's (1995) study of professional vision in oceanographers' work, he suggests that:

> the organization of perception is not ... located in the psychology of the individual brain and its associated cognitive processes, but is instead lodged within, and constituted through, situated endogenous practices. Such perception is a form of social organization in its own right.
>
> (Goodwin, 1995: 256)

This view of professional vision is consonant with Daston's (2008) emphasis on collective visual practices and competencies in scientific communities, and with Ingold's (2000) conception of perception as a distributed capacity: 'Perception is not an "inside-the-head" operation, performed upon the raw material of sensation, but takes place in circuits that cross-cut the boundaries between brain, body and world', Ingold (2000: 244) argues.

In addition, there is an innate relationship between perception and action (Goodwin, 1995: 256); perception is shaped by the technologies used in the oceanographer's work but perception is also shaping the action of the researchers. There is thus an intricate relationship between technology, perception and action. Professional vision is therefore what is always already distributed; perception is *per se* what is collectively determined (Lynch, 1985), the technology is *per se* a manifestation of underlying propositions (Bachelard, 1984), and action is the outcome from the mutual adjustment between technologies and perception. To further illustrate Goodwin's argument, Prasad's (2005) study of the use of advanced medical equipment in health care shows that medical doctors are developing what Prasad calls ' cyborg vision', that is, a form of professional vision that is inextricably entangled with advanced computer-based technologies. Prasad (2005) writes:

> In the cyborg visuality regime, images have become bits of data in cyberspace that can be, and are, manipulated by human beings. This does not mean that within this new visual regime, claims towards realism of images are disbanded. If that were so, there would be no reason to have MRI [Magnetic Resonance Imaging] radiological analyses. Cyborg visuality produced by MRI works within different frameworks of realism that does not seek mechanical reproduction of

the observed objects(s). MR images produce different reconfigurations of the body, each of which provide a partial perspective of the body and together they constitute the MR radiological gaze.

(Prasad, 2005: 310)

The magnetic resonance imaging does not offer self-explanatory images of the body. Since the body is already multiple (Mol, 2002), that is, subject to individual interpretations and professional vision in different medical sub-disciplines, the users of MRI technology must learn to translate the 'bits of data' into a credible account of the health condition of the patient. MRI offers specific images and forms of visuality that are embedded in advanced medical technologies, professional vision and specialized articulations. Prasad (2005) shows that sophisticated contemporary technoscientific health-care practices remain dependent on credible professional visual inspection, a specific form of gaze, to make effective use of, for instance, the MRI technology.

Studies of architects' work suggest – as we can recall from Chapter 3 – that artefacts play a central role in the professional vision of architects (Ewenstein and Whyte, 2007a, 2007b). Since aesthetic knowledge and aesthetic reflexivity is based on rather abstract or 'non-conceptual' skills, much architecture work revolves around the use of visual tools or artefacts such as images, photos, sketches, models and so forth. Expressed differently, the professional vision of the architect is often anchored in the ability to examine and account for images, sketches and models that are either collected from magazines, journals or internet sites, or are produced in-house (Yaneva, 2005). The non-conceptual aesthetic knowledge is manifested in sketches and models, thereby creating opportunities for communicating abstract ideas and what is 'in-the-making', not yet actualized but still in a state of becoming. Architects are relying on what Daston and Galison (2007) call 'trained judgement', the ability to interpret the various stakeholders' demands and wishes into a functional and credible solution, and the visual artefacts are playing a central role in bridging professional vision, aesthetic knowledge and qualified articulations.

In summary, while vision may, intuitively, appear as what is inherently unproblematic, historical accounts suggest that the 'image of vision' has changed significantly over the course of history (Crary, 1990, 1995, 1999). Rather than offering unproblematic images of reality, vision and visuality is instead shaped and formed by subjective capacities and dispositions. Furthermore, in a Lacanian tradition of thinking, vision is already infested by the other, that is, instituted norms, beliefs and assumptions, and therefore there is a split between what the eye actually perceives and what the observer experiences. The gaze is always the gaze of the other, Lacan contends. Further operationalizing this idea, what Goodwin (1994, 1995) calls professional vision is the entrenched visual capacity of a member of a professional community sharing a variety of skills and capacities. Professional vision is a form of collective visual capacity that often – but not always – is emerging in the integration and interaction of technologies, media, perception and action. Seeing is then mediated and entangled with action and practices. In the following, the professional vision of architects, the architect's gaze, is empirically examined. Here, it is suggested that architects in their everyday work draw on a register of specialized knowledge and their ability to inspect, perceive and apprehend social spaces and surfaces in a particular manner. The architect's gaze is thus a concept that brings together both knowledge and procedures of visuality into one seamless web of activities.

Introducing Brown Architects

Brown Architects, a major Scandinavian architect office, has its main office in Gothenburg, Sweden, but is located in a number of places in Sweden and Denmark. Brown Architects is nationally renowned and is a prestigious office, attracting some of the most skilled and creative architects, interior decorators, light designers, designers and engineers in Scandinavia. The office works in a wide variety of sectors pertaining to the built environment including housing, health-care buildings, schools, landscape design and architecture, furniture design and interior decoration. Brown Architects was founded in 1951; it has grown organically and through mergers and acquisitions, and is today one of the largest architects' offices in Scandinavia. The firm has ten offices and employs about 500 people, 99 of whom are partners of the firm. In 2001, Brown Architects won the prestigious national architecture award, the Kasper Sahlin Prize, for its Stockholm office.

The architect's gaze: seeing and knowing

One of the principal questions for the community of architects is to define what are their distinctive skills and competencies, unique features of the profession and their day-to-day practices. The general attitude among the interviewees was that architectural skills reside in the ability to bridge the engineering disciplines and aesthetic competencies. One of the interviewees emphasizes this concern for the community of architects:

> It is complicated to categorize architectural knowledge. What is it really? Here, we still have a problem to define what architects do, what kind of specific knowledge we can offer. This is an attempt to connect people and tacit knowledge and articulate it in discussions and seminars.
>
> (Interviewee 1)

One of his colleagues claimed that architects used 'intuitive knowledge':

> The architect profession is a bit special ... combining both technical and artistic features ... The classical view is that the more clear-cut engineering position uses knowledge that is verified, while for us, in our profession, there are components of intuitive knowledge or knowledge that are entrenched or trained. I do not know if this is true but it is the classic view ... Nevertheless, it is an issue in the industry as such because I think we have rather poor mechanisms for organized knowledge sharing.
>
> (Interviewee 4)

For the more critical interviewees, a number of persistent beliefs and assumptions regarding the nature of the architect's work inhibited more thorough knowledge-sharing procedures. For instance, one of the younger architects pointed at the organizational routines as the bearers of knowledge: 'It is tricky to speak of knowledge *per se* when talking about architects' work because it is just as much about skills and intuition. The process and the work procedures matter just as much' (Interviewee 2). As a consequence, he thought that 'the most effective knowledge transfer is still to work together and to sit down and listen to one another speaking' (Interviewee 2). The interviewee thus advocated a 'learning-by-doing' approach to the work and to knowledge-sharing. In general, the architects claimed that the work procedures were widely shared between teams and individuals, were very straightforward, and were in general easily understood. Yet the architect's work included certain components and events that were not easily translated into the communal vocabulary. Such tacit components of the work remain complicated to communicate both between co-workers and external stakeholders.

The architect's work procedures normally begin with a formulation of a *programme*. The programme includes the specifications of the future building, and clarifies the clients demand and needs. The architects start the process by contemplating and critically examining the possible solutions within the programme. One of the architects argued:

> The challenge is to identify the principal problems within the programme and give as good answers to them as possible. Then you can start the sketching process and the programme analysis straight away. In many cases, projects are initiated as a rather fuzzy question or in terms of a continuation of a previous project.
>
> (Interviewee 3)

Rather than beginning to sketch possible solutions from the outset, many of the interviewees argued that they needed some time to consider various alternatives and to have the chance to avoid applying ready-made solutions to problems. This stage in the process included the use of various materials such as photographs and images. In addition, activities such as visiting the site was of great importance for the forthcoming work:

> Once you've visited the site and have been thinking for a while, you may contrive of entire new solutions ... There needs to be a period of time where you juggle different alternatives ... One mustn't follow the first instinct.
>
> (Interviewee 3)

In general, architects take great pride in offering as aesthetically qualified solutions to perceived problems as possible, and therefore they maintain that it is important to 'exceed the expectation of the client' and to provide what had not been anticipated. One of the architects said:

> In the sketching method there is an implicit assumption that it is all about identifying the unique features of every assignment on the basis of the conditions that we operate under. To create what no one could really anticipate.
>
> (Interviewee 4)

After the initial phase of reflection and problem definition, the architect translates the programme into some kind of visual space, a rough sketch of how the building may be materialized. This stage of the process is referred to as the *sketching*, the use of various media – ranging from pen and paper to advanced three-dimensional (3D) computer-aided design (CAD) applications – to develop the features of the building. Some of the architects thought it was important to keep as many opportunities open as possible in the process but that posed a challenge when working with clients very eager to see the actual architecture taking a more fixed form:

> One important aspect is to avoid fixing some of the parameters, to be capable of maintaining an openness as long as possible. That is accomplished not by designing objects and entities and giving them a form. For some clients, this is quite complicated to handle.
>
> (Interviewee 2)

The sketching phase is in many ways central to the work procedure. Here many visual artefacts are used to enable the communication of aesthetic and functional solutions. For instance, the use of photographs from architecture journals or architecture office homepages are widely used in the process. One of the architects emphasizes the use of pictures:

> Quite often, the picture is important. You do the sketching and try to visualize and then you make use of reference pictures. You bring photos

and things like that from previous projects. Even in the presentation of one's own project, you make use of photos to capture a certain attitude or feeling or a possible solution for something.

(Interviewee 2)

Later on, the use of 'reference photos' were addressed in more detail:

Q: How do you discuss architecture?
A: A lot of talk derives from reference photos, reference projects; when sketching something or building a model, more often than merely speaking about it. At times you need to help clarifying – a first sketch could be quite vague. You need to explain it: 'this is what I would like to accomplish', but to just speak about it is quite complicated.

(Interviewee 2)

Another interviewee also underlined the need for having visual artefacts or tools to support the verbal communication, but also argued that photographs are deceiving in terms of seducing the architect to adopt previous solutions rather than engaging more fruitfully with possibilities of the programme and the actual site:

You need to build on your experience, on photos you find in the literature, or on excursions. At the same time, it is damn easy to adopt this kind of information so you need to be careful. The foundation for a good solution is in one way or another already present in the very programme.

(Interviewee 3)

In other words, there is a clash between the architect's ideology of 'producing the new and previously unseen' and the exposure to images and photographs of previous contributions.

Further down the process, when the sketching phase has resulted in a more tangible solution, architects in many cases (but not always) use models to enable further communication both with colleagues and stakeholders such as clients and end-users. While the sketching represents the first stages where the programme is materializing into actual solutions, the model is capable of adding spatial dimensions; the sketch is a two-dimensional (2D) tool and the model is a 3D tool. One of the interviewees emphasizes the need for using various complementing media:

You don't produce architecture through 'cognitive processes' [in Swedish: *Man tänker ju inte fram arkitekuren*]; it is not a logical consequence; there is no single correct answer to an articulated problem but it is instead a process of testing and decision-making ... The more dimensions you add and the more you can visualize, the faster you can advance. Therefore, the use of models adds a dimension to the sketch.

(Interviewee 3)

Producing qualitative architecture is the ability to balance and combine practical, material, aesthetic and financial concerns, and to negotiate various solutions with stakeholders. This complicated balancing of interests and concerns demands visual tools that can enable communication. 'A model is the best item to collaborate around, quite simply' (Interviewee 3), the architect contended. One of her colleagues emphasizes the same features of models: 'I think the most important feature of the physical model is that more people can participate in the process at the same time, in one way or another' (Interviewee 2). One of the interior decorators agreed that the 2D visual tools are impotent in capturing the full complexity of the subject matter: 'You cannot just work with CAD. CAD is certainly not the adequate tool when working with interior decoration. You can integrate things but to produce models is complicated' (Interviewee 6).

The enacted work procedures strongly emphasize the ability to move gradually from rather abstract propositional formulations of the programme to a more material and tangible artefact. The architect's competence to translate the programme and its specifications into actual solutions and to take into account the specific features of the actual site is a process that is relying on what here has been called 'the architect's gaze', that is, the capacity of conceiving of a forthcoming building on the basis of visual inspection and ways of seeing. The architect examines the programme, visits the site, searches for and explores reference photographs and images, works with sketches, builds models and finally provides blueprints when all aspects of the building are negotiated with clients and other stakeholders. The architect's gaze is, therefore, of central importance for the entire process and, consequently, for the professional identity *per se*. One common method to reproduce and 'calibrate' such a shared 'architect's gaze' collectively is to undertake excursions, travels to visit built architecture and environments on-site. This tradition was strongly advocated by the interviewees and was generally regarded of central importance for the day-to-day work and for professional identity. One of the interviewees argued:

To visit the built architecture is of great importance. Then you may observe how the assignment is handled. If it is a church or a housing project, then you examine how the building fits into the city planning and down to the details and what materials are used. For most of us, that is a way of identifying solutions to problems; a form of knowledge about how to handle something.

(Interviewee 1)

The actual visiting of built architecture gives the architect the opportunity to explore features of the building that are not captured by other visual tools. When visiting buildings, one 'apprehends the scale and the space itself; which you cannot do with a picture. In addition, the context [is examined]' (Interviewee 1). Another interviewee expressed a similar belief:

> First, it is professionally rewarding to visit a place ... Architecture *is* built environment. You cannot fully experience architecture unless you visit the site. You need to ... capture it to be able to explore all parts. That is why it's important to go in these excursions and not only consume images in journals.
>
> (Interviewee 3)

Another architect pointed to the 'immersive' experiences when visiting a site, speaking about 'layers' of insight:

> You acquire knowledge on a variety of layers. That's why it is such an effective method. Including everything from the formulation of the assignment [in Swedish: *förhållningsätt till uppgift*], how a colleague has interpreted the assignment, explored the opportunities of the site, and how the programme and the demands have been interpreted. Of course, the design and the form, choice of materials and so forth. On all layers, you may learn new things by visiting finished projects.
>
> (Interviewee 4)

While photographs and other images of buildings are excellent tools in conveying meaning or aesthetic objectives, they can never fully capture the brute immediacy of an actual building. The 2D medium is never capable of apprehending the actual sense of the building:

> We work quite a lot with two-dimensional tools ... The experience of being part of the room is very, very important. To understand what one is sketching, to spend time in environments where you can sense how large a space is ... Then you need to visit places. That's how the brain works. You cannot just look at a sketch. When you have worked for 25 years, then you create an understanding [in Swedish: *förståelse*] but never a sense [in Swedish: *känsla*] of the space. It is imperative to spend some time on the sites.
>
> (Interviewee 5)

The architects also noted that the practice of conducting excursions was no longer exclusive to the community of architects. Today, a great number of construction industry actors invest time and resources in visiting buildings and other built environments on site:

> In one way, that is the main way for professional development among architects of all ages. I would like to claim that it was the architects who 'invented' that method for knowledge development. The first architects went to Italy or France to get new impressions and ideas. That has been the case throughout history. Today, it is not only architects doing excursions but the whole industry travels to visit places. We start to think that it has been part of our activities and that we are bypassed by the clients who are out there all the time and are more informed than we are.
>
> (Interviewee 4)

The excursions, organized at the same time every year, served as a common reference point for the collective work in the office. Although different groups (e.g. architects, landscape architects, light designers, interior decorators or furniture designers) had the opportunity to visit individually chosen excursions (for instance, the designers occasionally visited the annual furniture exposition in Milan), experiences, insights and expressions from the excursions were presented in-house during seminars and through the circulation of photographs and other documentation. Excursions then served the important role of helping to constitute a shared sense of architectural spaces and materials through joint experiences. Besides the latent function of the excursions (i.e. the 'bonding' within teams, departments or between the firm's different geographically distributed offices), the excursions had the manifest function of providing a shared ground for continual discussion on architecture and built environments. More specifically, the excursions played a central role in anchoring highly distributed and at times rather abstract competencies and skills in a shared experience engaging all senses.

In summary, architects share a certain professional vision and, to further support this collective capacity of seeing, various visual tools are used. In order to 'calibrate' the professional vision, architects undertake excursions where actual built spaces are experienced on site. Buildings are here examined in their full presence, in their materiality and in their location. Built spaces are perceived from a variety of perspectives and using a number of senses.

The architect's gaze and the embodiment of knowledge

The architect's gaze qua collectively accomplished skill and competence is what resides in joint and immersive experiences. Visiting buildings on site is therefore central for the architect's capacity to bridge the abstract and the concrete. In their work, architects move from written documents (i.e. the programme) to spatial solutions to perceived problems. As the process emerges, the forthcoming building is gradually being materialized in sketches, CAD images, physical models and the final blueprint. This gradual materialization and stabilization is a social process relying heavily on the architect's professional vision, the ability to perceive and sense the opportunities of an actual site. Throughout the process, there is a dynamic inter-relationship between such a professional vision, enacted vocabularies, written documents, and visual artefacts, such as photographs or models. However, the architect's gaze is never external to the individual architect; instead, the gaze is embodied in the perceptual capacities of the architect in the form of a particular 'way of seeing' and examining a particular site or programme. That is, the architect's gaze is a form of professional vision that in one single process integrates a series of skills, experiences, ideologies, norms and values and conceives of credible solutions to perceived problems. Speaking of knowledge and know-how in architecture, the architect's gaze is what in one single instant embodies the totality of entrenched individual or collective skills and experiences. The architect's gaze is thus a gaze in the Lacanian sense of the term, that is, the gaze of 'the other'.

Architectural knowledge is formally trained in university programmes, and thereafter further refined and developed in actual professional work. Architectural knowledge is also manifested in the various visual artefacts and tools, such as photographs used in the day-to-day work. It is part of the architect's training and acquired professional skill to have the ability to inspect and determine visually whether a photograph of a building is worthy of attention. Expressed differently, the gaze of the architect is always penetrated by what are regarded legitimate knowledge and ideologies within a specific field. Perception, and perhaps more adequately, what Daston and Galison (2007: 368) calls 'practices of seeing', are an integral part of the architect's work; architects use their professional vision in every single part of their day-to-day work. Architects look at, examine, review and browse through a great variety of visual artefacts on an everyday basis; their work is grounded in a particular professional vision.

Such a professional vision does not come for free but is instead what is acquired through long-term engagement with architectural issues and concerns, including the discussions with colleagues and peers and by investing time in keeping track of the latest changes and events in the field. From an organizational perspective, the professional vision, the architect's gaze, is a form of organizational capability that must be actively supported and managed. For instance, that is why excursions, seminars and workshops are organized at Brown Architects. When collectively examining and debating built environments, the professional vision is reproduced and further developed. The ability to apprehend, perceive and articulate experiences from social spaces is at the very heart of the profession. Architects complained about the time compression of the projects and the increasing pressure to cut costs in the construction industry. The ability to share a professional vision, at times even operating without taking the detour to verbal articulation, saves time and effort. The architect's gaze is then what is not only an individual skill but equally a collective or organizational skill.

However, the architect's gaze, the gaze of the other, is also subject to suspicion among architects. Some of the architects in the study articulated their concern that they tended to adopt ready-made solutions to perceived problems and that their professional vision may be skewed by clichés and 'off-the-shelf solutions'. The professional vision is here interpenetrated by fashions that as such are violating the predominant ideology of creating the unique and new in every single project. However, following a Lacanian frame of reference, this sense of always having one's perception already shaped and formed by social forces beyond one's control is part of the human condition in contemporary society. The anxiety over one's authenticity of vision is a perennial issue – are there any substantial opportunities for 'immaculate perception' non-biased by previous experience and discourses? (Denzin, 1995). Therefore, architects strongly emphasize the need for perceiving social spaces at site, in its actual environment, but they do at the same time believe they have to avoid 'becoming boring' or 'predictable' by actively putting into questions one's own thinking. 'One mustn't follow the first intuitive thought', one of the architects claimed.

Speaking about the use of knowledge, know-how and other forms of intellectual resources in architecture work, the professional vision of the community of architects is central to their understanding from both a practical and a theoretical perspective. Everyday practices in architecture offices and bureaus are strongly shaped by the professional vision, the architect's gaze. From a theoretical perspective, the corpus of knowledge-management literature has been only marginally concerned with examining knowledge that is embodied or rather knowledge that resides in the intersection between cognitive and perceptual capacities, that is, in the ability to combine seeing and thinking effectively in everyday practice. In addition, the concept of professional vision and the gaze strongly emphasizes the collective and joint accomplishment of establishing a calibrated and widely shared mode of inspection or way of seeing. The gaze of the architect is thus always the gaze of the collectivity. Speaking of knowledge in such distributed and social terms opens up new opportunities when examining the concept of knowledge in its actual use. This chapter thus contributes to a theoretical perspective on knowledge qua organization resource that underlines the relational and distributed nature of knowledge, that is, a non-foundational (Rorty, 1998) and constructivist view of knowledge (Tsoukas, 1996; Herrstein Smith, 2005). More specifically, this chapter contributes to the emerging literature examining the intersection between human perception (i.e. tactile, visual, auditory, olfactory competencies) and the faculty of knowledge. In this view, knowledge does not solely reside in the cognitive capacities of human beings, but is instead what is distributed over a variety of human skills and competencies.

Summary and conclusion

The accumulated individual and collective know-how, skills and experiences of architects do not simply reside in cognitive human capacities, but are instead distributed over a variety of resources and assets, such as work procedures, visual artefacts, such as photographs and full-scale models, jointly shared vocabularies and so forth. In addition, the totality of the architect's acquired know-how is present in the professional vision of the architect – the architect's gaze. The architect's gaze helps in interpreting a specific assignment (i.e. the programme), examines the opportunities and limitations of a certain site, inspects and analyzes relevant photographs and computer-generated images, and evaluates various alternatives being produced in the sketching process. Rather than being external to the knowledge work, the professional vision of the practising architect is what shape, form and frame the activities constituting the knowledge-intensive work that ultimately provides the blueprint for the forthcoming building. The knowledge-intensive work of the community of architects is thus a form of aesthetic work embedded in perceptual capacities and in the ability to communicate meaning verbally and symbolically. Therefore, it is complicated to speak of knowledge-intensive work as being strictly cognitive. Instead, such work is always of necessity emerging in a network of integrated and inter-related resources, skills and activities. Knowledge-intensive work is then an assemblage comprising a variety of resources that are brought together and mobilized in the pursuit of formulating a credible set of solutions to a range of identified and articulated problems. Speaking of knowledge management as a set of theoretical propositions, concepts and models, the study of Brown Architects shows that knowledge management practice is essentially embodied and anchored in the perceptual capacities of the practising architect. Just like a laboratory researcher learns to listen to their machinery (Mody, 2005) and a surgeon is exposed to haptic sensations derived from the touching of the patient's body (Prentice, 2005; Johnson, 2007), so must the practising architect learn to look and articulate ideas as one single integrated process. In professional vision, looking and talking is always intimately related; in order to speak, one must look; looking is preceded by speech.

5   Knowing the concrete
Knowledge and skills in a specialist rock construction company

Introduction

The three preceding chapters addressed how to support site managers in their day-to-day work, and examined how architects are dependent on external resources and communication and their visual capabilities in their work. In this chapter, a study of the knowledge-sharing practices in a specialized construction company will be reported. The company in question, here referred to under the synonym ConCo (short for the Concrete Company), operates in a niche of the construction industry specializing in rock construction work of various kinds. Among the key competencies of ConCo are spray-concrete construction work, rock injection and rock reinforcement. Such expertise is used primarily in construction work that is infrastructural, laid down in technical systems that are widely taken for granted and unattended to unless they fail to support their function. In the following chapter, two complementary aspects of the knowledge work in ConCo will be examined.

First, in the analysis of how ConCo's site managers and other significant decision-makers manage and lead their daily work in the construction sites, the concept of social capital is invoked. 'Social capital' is a complicated sociological term denoting the value of social relations in a particular field. Other forms of capital (e.g. financial or human capital) are essentially possible to locate to specific organizations or individuals, but social capital is, by definition, what resides in the *relations* between human beings. In the analysis, social capital is what supports and anchors knowledge-sharing procedures in the site managers' field within ConCo. Since they tend to be unwilling to invest time and effort in formal knowledge codification procedures and the work in ConCo is riddled by ambiguities and unanticipated occurrences, site managers essentially rely on a network of expertise they can consult when dealing with emerging problems and challenges. The social capital is, in other words, the network of relations that enable knowledge to circulate and be distributed between individuals, construction projects firms, and so forth. The concept of social capital enables an understanding of how knowledge is collectively shared in communities that are based on oral communication.

Second, the concept of aesthetic knowledge addressed in the literature on organization and aesthetics will be invoked to examine how the line of demarcation between skilful and less skilful operators in ConCo is drawn. While construction industry co-workers such as engineers, site managers, foremen and operators favour a practice-based and straightforward vocabulary when addressing their work, some domains of expertise defy such language devoid of ambiguities and uncertainty. For instance, when site managers express how some of their co-workers are capable of undertaking certain assignments, they draw on a vocabulary that shares many qualities with what has been referred to as aesthetic knowledge. In the second half of the chapter, the connections between rock construction work and aesthetics will be addressed in more detail.

Social capital in rock construction work

Social capital is conceived of as the shared and mutual trust that individuals develop in their joint collaborations. Rather than assuming that individuals working in organizations and over organizational boundaries are operating as autonomous and detached entities, devoid of interaction and communication, social capital underlines the collaborative and collective nature of social existence. In a knowledge-management setting, individual organization members are at times seen as separated cognitive universes where skills, know-how and capabilities are located. In other cases, knowledge is fundamentally conceived of as a 'social accomplishment' (Orlikowski, 2002), the outcome from a collective effort embedded in communicative skills and the capacity to take the role of the other. In the construction industry, most operations and activities are structured around the participants' ability to collaborate and continuously exchange information. Rather than being an industry nourishing an ideology underlining individual contributions (as for instance, in some domains of scientific work), the construction industry is from the outset based on the ability to manage and organize complex informational, material and design-based interchanges. In construction projects – in many cases the smallest operative unity of analysis – diverse and complex skills in operations management, design management, engineering and craft skills are brought together. The ability to coordinate these diverse skills is one of the principal capabilities of the acting manager. This domain of expertise in ConCo includes construction techniques such as roof-bolting, shotcreting and rock reinforcement, in many cases under harsh conditions in tunnels and other forms of subterranean constructions, such as subway systems or sewage systems. Rather than pursuing advanced forms of knowledge sharing, that is, knowledge that is formalized and mediated (i.e. stored and distributed in computer systems), ConCo is essentially relying on an oral culture, and the social capital developed and nourished in the firm and the construction industry to circulate and share knowledge. Contrary to other industries (e.g. technology-based companies, pharmaceutical companies), construction firms pay relatively little attention to the formalization of knowledge. Instead, word-of-mouth and personal contacts play a central role in the 'knowledge-system' (Dougherty, 2007), dominating in construction companies such as ConCo (Bresnen, Edelman, Newell *et al.*, 2005). However, rather than assuming that this is an 'archaic' or outmoded form of knowledge-sharing, it may be, as the chapter suggests, that this embedding of expertise know-how in social capital and oral communication is the most effective way to store and distribute knowledge in an industry catering for a wide range of services. Seen in this view, social capital is of central importance for ConCo and other construction firms, and is also of great interest for the field of knowledge management studies.

Social capital and the sharing of knowledge

Social capital is a central construct in contemporary sociology and one of the most debated concepts (Coleman, 1988; Portes, 1998). Coleman (1988: S98) emphasizes that social capital is 'defined by its function'; it is not 'a single entity' but a 'variety of different entities', with two elements in common: 'they all consist of some aspect of social structures, and they facilitate certain actions of actors – whether persons or corporate actors – within the structure'. Social capital is therefore, like other forms of capital, 'productive'; it is making possible 'the achievement of certain ends that in its absence would not be possible', Coleman (1988: S98) contends. Coleman (1988) contrasts physical, human and social capital in terms of how tangible these sources of capital are. While physical capital is 'wholly tangible' and created by changes in materials to form tools that 'facilitate production', and human capital 'less tangible' and created by changes in persons that 'bring about skills and capabilities that make them able to act in new ways' (Coleman, 1988: S100), social capital is yet less tangible for it exists in 'the relations among persons': 'Unlike other forms of capital, social capital inheres in the structure of relations between actors and among actors', Coleman (1988: S98) concludes.

Pierre Bourdieu develops an entire theory about how economic, cultural and social capital structure what he calls *fields*, delimited social domains characterized by individual rules and regulatory mechanisms. He defines social capital as 'the aggregate of the actual or potential resources which are linked to possession of a durable network of more or less institutionalized relationships of mutual acquaintances or recognitions' (Bourdieu, 1985, cited in Portes, 1998: 3). In his review article on social capital, Portes (1998: 3) argues that Bourdieu's analysis is '[a]rguably the most theoretically refined among those that introduced the term in contemporary sociological theory'. Contrary to other forms of capital, controlled by the individual, social capital is by definition distributed; it 'inheres in the structure of their relationships' in Portes's (1998: 7) formulation. That is, in order to possess social capital, the individual actor must relate to others and it is those relations with others that are the actual source of advantage in a social system. 'Definitions of social capital share in common', Bresnen, Edelman, Newell *et al.* (2005: 236) write, 'the key idea that the networks of relationships in which people are embedded act as an important resource and thus a source of competitive advantage to the firm'. Maurer and Ebers (2007) speak of social capital in similar terms:

> [T]he core intuition behind the notion [social capital] is that it signifies an asset available to individual or collective actors that draws on their actors' positions in a social network and/or the content of these actors' social relations.
>
> (Maurer and Ebers, 2007: 262)

More specifically, social capital is valuable for the focal firm because 'social capital has potential value because it provides an opportunity for actors to access information and resources in their network', Maurer and Ebers (2007: 262) suggest. Even though social capital is distributed across communities, individuals can benefit greatly from their access to social capital. Burt (1997: 339) suggests that 'managers with more social capital get higher returns to their human capital because they are positioned to identify and develop more rewarding opportunities'. Moreover, Burt (1997: 349–50) found a significant correlation between early promotion of managers and received bonuses and social capital. Finally, which is of great importance for the site manager position, Burt (1997: 355) found evidence suggesting that 'social capital is less valuable to managers with numerous peers ... Second, social capital is more valuable to boundary managers'. Since site managers tend to conceive of themselves as being by and large on their own and since they are hubs in extensive networks of relations, social capital is potentially a very valuable resource in their day-to-day work.

Although sociologists and other proponents of a social capital perspective tend to strongly emphasize the positive consequences of social capital, there are also negative consequences associated with the term. For instance, conformism and so-called free-rider problems are observable in communities in control of substantial social capital. For instance, Smith (2002: 527) demonstrates that 'blacks receive comparatively lower returns than whites to their human capital investments, even when they occupy similar levels of authority and are located in the same industries'. One explanation of this inequality is that the relatively lower degree of social capital in the black community generates a lower economic return on investment in human capital. Portes (1998: 8) also emphasizes that such communities may demonstrate what sociologists has came to call 'bounded solidarity', a solidarity reserved for their peers and in a rather narrow social community.

Mouw (2006) here speaks of the tendency of 'similar people to become friends with each other' as 'social homophily'. Speaking of social capital in a construction firm, Bresnen, Edelman, Newell *et al.* (2005: 237) point out that relying on social capital when sharing knowledge is very costly and inefficient since it involves 'considerable effort' in establishing and maintaining relationships. Moreover, Maurer and Ebers (2007: 275) found in their study of German biotechnology start-ups that there were two identifiable limitations derived from the reliance on social capital. First, Maurer and Ebers (2006) speak of 'relational lock-in' in situations where firms were unable to renew the basis for their social capital as the start-ups evolved. Second, Maurer and Ebers (2007) found evidence of 'cognitive lock-in' in cases where the biotechnology firms lacked access to additional viewpoints. As a consequence, the biotechnology entrepreneurs '[l]acked the capacity and competence to relate effectively to partners outside the scientific community', Maurer and Ebers (2007: 277) note. The ability to bring in alternative viewpoints thus led to what Janis (1982) calls 'groupthink', a convergence in thinking that is potentially harmful for the company.

However, taken together, the concept of social capital is useful when explaining how knowledge is developed and shared in various social communities. A stream of research emphasizes that social capital is being mobilized in innovative and creative knowledge-intensive communities (for an overview, see Adler and Kwon, 2002; Willem and Scarbrough, 2006). The importance of social capital in knowledge-intensive work has been emphasized in a number of studies (Nahapiet and Ghosal, 1998; Yli-Renka, Autio and Sapienza, 2001; Newell, Tansley and Huang, 2004; Inkpen and Tsang, 2005). In this setting, Subramaniam and Youndt (2005: 451) define social capital as '[t]he knowledge embedded within, available through, and utilized by interactions among individuals and their networks of interrelationships'. Nahapiet and Ghosal (1998: 243) distinguish between the structural, the relational and the cognitive dimensions of social capital. Structural social capital is embedded in practices and institutions. Relational social capital is capital that is jointly developed and utilized when individuals collaborate in a specific field. Cognitive social capital, finally, is the individual co-workers' perceived value of being part of a larger community, such as a professional group, a project team, or a department sharing their know-how and expertise. Nahapiet and Ghosal (1998: 245) emphasize that the function of social capital is to 'increase the efficiency of action'. Subramaniam and Youndt (2005: 451) found that investment in what they refer to as organizational and human capital is only poorly exploited unless being accompanied by investment in social capital. They argue:

> To effectively leverage investments in human capital, it may be imperative for organizations to invest in the development of social capital to provide the necessary conduits for their core knowledge workers to network and share their expertise. Organizations that neglect the social

side of individual skills and inputs and do not create synergies between their human and social capital are unlikely to realize the potential of their employees to enhance organizational innovative capabilities.

(Subramaniam and Youndt, 2005: 459)

Cohen (2007: 251) argues that the essential elements of 'social capital creation' is: (1) to provide time and space to meet and work closely together so they can 'develop mutual understanding and trust'; (2) to build trust by demonstrating trustworthiness and delegate responsibilities; (3) to ensure equality in terms of opportunities and rewards to foster 'commitment and cooperation'; and (4) to examine existing social networks to see where valuable relationships can be preserved and strengthened. In Cohen's (2007) view, social capital is an organizational resource that can be deliberately managed and elaborated upon. It is then no wonder that the importance of social capital has also been emphasized in leadership work. Balkundi and Kilduff (2005: 953) advocate what they call a 'social network approach' to leadership and here '[l]eader effectiveness involves building social capital that benefits individuals in the organization and extending the social networks of subordinates to facilitate career networks and leadership potential of subordinates'. The management of knowledge-intensive work is therefore somewhat paradoxical; on the one hand, knowledge-intensive work is bound up with individual skills and expertise while, on the other hand, all knowledge is the product of social interaction and collaborations. Speaking in terms of distributed innovation between organizations, it is possible that social capital will play a more accentuated role in bringing heterogeneous interests together, making them operate functionally.

Heedful inter-relation and social capital

One specific domain of research drawing on a social capital perspective is a series of studies of what Weick and Roberts call 'heedful interrelation'. Weick and Roberts (1993), making reference to the Oxford philosopher Gilbert Ryle (1949), speak of organizations as 'collective minds' where different individuals' thinking and acting are inter-related in what they refer to as a 'heedful' manner (Weick and Sutcliffe, 2006). '[H]eedful performance is', say Weick and (Roberts, 1993: 362), 'the outcome of training and experience that weave together thinking, feeling, and willing'. Heedful inter-relating, the 'dispositions to act with attentiveness, alertness, and care' (Weick and Roberts, 1993: 374), is a central competence for organizations operating with tight couplings and complex inter-relations; for instance, in new product-development teams, operators in environments dealing with riskful activities, such as nuclear plants' control rooms, or teams of surgeons and nurses. Heedful inter-relating is of central importance because it enables thoughtful and effective communication, and the sharing of information in the course of action. When heed declines, performances become '[h]eedless,

careless, unmindful, thoughtless, unconcerned, indifferent. Heedless performance suggests a failure of intelligence rather than a failure of knowledge. It
is a failure to see, to take note of, to be attentative to' (Weick and Roberts,
1993: 362). For Weick and Roberts (1993), heedfulness is a form of social
capital making better use of the know-how that exists in a team. For instance,
a surgeon that can interact heedfully with a nurse can eliminate unnecessary
communication, avoid misunderstanding, and function more smoothly as
a team (Edmonson, 2003). One of the aspects of heedful inter-relating is
the ability to 'cultivate care' in teams (von Krogh, 1998; Styhre, Roth and
Ingelgård, 2002). Weick and Roberts (1993) argue that care is a quality
that is demonstrated *in actu*, in the very undertaking of a practice, rather
than existing prior to action: 'Care is not cultivated apart from action. It
is expressed in action and through action. Thus people can't *be* careful,
they *are* careful (or careless). The care is in the action' (Weick and Roberts,
1993: 373). Weick and Roberts emphasize the importance of understanding
social interaction as what is embedded in the ability to 'take the role of
the other' (Mead, 1934) and to intersubjectively anticipate forthcoming
actions in the unfolding of a sequence of joint actions. Weick and Roberts
thus conceive of complex and tightly coupled activities as being distributed
between a number of individuals and resources, brought together by the
capacity to inter-relate heedfully. More recently, Druskat and Pescosolido
(2002) used the concept of heedful inter-relating in a study of team-based
work and Dougherty and Takacs (2004) examined heedfulness in innovation
work. Druskat and Pescosolido (2002: 293) summarize the argument:

> [H]eed is not behavior; rather it refers to the way in which behav-
> iors are enacted. Interpersonal interactions assembled with heed are
> attentive, purposeful, conscientious and considerate. They increase
> team effectiveness by improving members' ability to work together
> effectively ... Without the enactment of heed, interpersonal interactions
> and relationships are paid little regard.

Dougherty and Takacs (2004: 574) say: '[H]eedful interrelating is a basic
theory of effective social relationships that applies more generally to non-
efficiency organizations, such as innovative organizations'. They examine
successful innovation work as a form of systematic play embedded in heedful
inter-relating. Weick and Roberts (1993) warn that heedful inter-relating is
a capacity or competence in organizations that can easily dissolve unless it
is supported by a culture affirmative of heedful inter-relating:

> A different way to state the point that mind is dependent on social skills is
> to argue that it is easier for systems to lose mind than to gain it. A culture
> that encourages individualism, survival of the fittest, macho heroics,
> and can-do reactions will often neglect practice of representation
> and subordination. Without representation and subordination,

comprehension reverts to one brain at a time. No matter how visionary or smart or forward-looking or aggressive that one brain may be, it is no match for conditions of interactive complexity. Cooperation is imperative for the development of mind.

(Weick and Roberts, 1993: 378)

Rather than assuming that heed is always already in place in organizations, in teams or in professional groups, one can think of it as what is continuously reproduced and re-instituted through actions. Heedful inter-relating is part of the social capital of the firm and, therefore, it needs to be nourished and supported to avoid being eroded.

In one of the few studies of the role of social capital in construction industry, Bresnen, Edelman, Newell *et al.* (2005: 240) found that co-workers in the organization studied used primarily 'very traditional means' for communication (i.e. direct contact, telephone and e-mail). More 'advanced mechanisms' for accessing a wider knowledge base, such as company intranets or internet access were not so well used. The co-workers favoured speaking directly to one another, preferably face-to-face. Social capital, therefore, played a central role in connecting the co-workers to one another. In general, social capital was of great value for the company but there was also a downside to it; what Bresnen, Edelman, Newell *et al.* (2005: 242) call 'within-group cohesion' promoted a somewhat 'inward-looking perspective' and created problems when there was a need to 'access new information or to harness different expertise to address novel problems'. Social capital may, therefore, '[create] powerful forces that can cancel out any beneficial value creation efforts', Bresnen, Edelman, Newell *et al.* (2005: 243) warn.

In summary, when adhering to a social capital perspective, social practices embedded in a culture or professional ideology emphasizing the value of heedful interaction and inter-relating, pay substantial attention to the collective developing and sharing of organizational competencies, know-how and skills. In organizations with high levels of social capital, co-workers are inter-relating heedfully and are willing to share their insights and know-how with one another. In such organizations, knowledge is not an individual and clearly bounded resource, but is collectively mobilized and used in everyday practice. At the same time, high degrees of social capital may prevent and inhibit new thinking and innovative ideas; conformism and group-thinking is potentially the price paid for shared social capital.

Social capital in site management work in the construction industry

The construction industry is based on advanced civil engineering competencies, project management skills, and the ability to lead and coordinate diverse professional and occupational groups in everyday work. Construction industry representatives on all managerial levels and in all occupational groups tend to appreciate what is regarded as the creative and innovative nature of the work, that is, the very construction of a unique building on the basis of blueprints and various materials. While other industries develop prototypes prior to full-scale production, the argument goes, there are no prototypes in the construction industry but only individual buildings constructed from scratch. At the same time as construction industry representatives declare and demonstrate their loyalty to the industry – many spend their entire careers in the industry or have inherited the job from their fathers – there is also evidence of harsh work conditions for many occupational groups. One group particularly exposed to stress, ambiguous roles and significant workload are the site managers, the project managers in charge of individual construction sites. Studies reported (see Chapter 2) by Djerbarni (1996), Mustapha and Naoum (1998), Fraser (2000), Davidson and Sutherland (2002) and Lingard and Francis (2004, 2006) show that site managers work long hours and experience an everyday work-life dominated by continuous disruptions and excessive workload. Faulkner's (2007) study of construction engineers emphasizes the heterogeneous skills demanded, ranging from 'analytical problem-solving' to 'presentation and communication skills' and the ability to handle delicate interpersonal situations. These skills are largely gained on the job, and include project management, accounting, team building, and the ability to build and maintain networks of contacts. Like perhaps no other engineering discipline, civil engineering in the construction industry is based on 'learning-by-doing':

> Above all, building design engineers, like other engineers, build up and rely on a huge body of cumulative experience of what 'works and what doesn't', from which they are able to make 'sound' judgments, not only about the design but also about the management and 'politics' of complex projects.
>
> (Faulkner, 2007: 335)

Studies of site management work shows that site managers tend to regard some parts of their work as being less rewarding, for instance administrative work and juridical issues to be addressed (Styhre, 2006). What site managers appreciate is the day-to-day work on the site, dealing with the practical challenges emerging in the construction process. These findings are in line with Faulkner's study, reporting that construction engineers lament the loss of 'real' engineering work (Faulkner, 2007: 345) and replaced by work assignments that are 'soft' and 'people-oriented'. A similar attitude is reported by Perlow (1999) in a study of software engineering work in a Fortune 500 firm:

> I found that engineers distinguished between 'real engineering' and 'everything else' that they did. They defined real engineering as analytical

thinking, mathematical modeling, and conceptualizing solutions. Real engineering was work that required using scientific principles and independent creativity. It was the technical component of the engineers' deliverables that utilized the skills that engineers acquired in schools ... In contrast, 'everything else' translated mostly to interactive activities as disruptions to their real engineering, although further examination revealed that interactive activities were critical to the completion of the engineers' jobs. When engineers confronted barriers, they often turned to other engineers for help.

(Perlow, 1999: 64)

Contrary to Faulkner's (2007) and Perlow's (1999) engineers, site managers do in general recognize their leadership role and regard it an integral part of their work. In addition to the loss of 'authentic' site management work, site managers perceive their role as being somewhat isolated from other site managers and from the line organization. Site-management work is very much the work of a 'lone wolf' trained to take care of his own responsibilities. The ideology of site managers strongly underlines the importance of carrying out the work on one's own and being able to be in charge of the most complex undertakings. At the same time, site managers are learning to develop and maintain a substantial network of colleagues, former colleagues, experts and so forth, being able to give good advice on how to handle various emerging situations and problems. Site managers are therefore both regarding themselves as being isolated while they can demonstrate a large number of what Granovetter (1973) calls 'weak ties' in the industry. Expressed differently, the social capital of the site manager is of central importance for the long-term performance.

Introducing ConCo

The study reported in this paper is a based on a single-case study of the medium-sized specialist construction company ConCo, located in Stockholm and Gothenburg in Sweden. ConCo specializes in underground and rock construction work and provides advanced construction services in rock and ground reinforcement, shotcreting and other uses of concrete in rock constructions such as tunnels or underground systems. ConCo is competing with the 'Big Three' Swedish construction firms (i.e. Skanska, NCC and PEAB) and a smaller number of small and medium-sized firms. More recently, European companies have started to operate on the Swedish market, including German, Austrian, Swiss and Norwegian companies. The co-workers at ConCo perceive the firm as a knowledge-intensive firm, running the slogan 'build on knowledge', capable of taking on any assignment pertaining to their domain of expertise.

Social capital in construction engineering work

The interviewees at ConCo argued that their work contained a substantial amount of expertise and insights into 'what worked and what did not'. Such expertise and insights were accomplished through practical work and learning-by-doing. Being able to listen to colleagues' stories and advice and learning to handle the small but significant events and occurrences were therefore of central importance for any effective site manager. Site managers also gained recognition and contributed through running construction projects effectively, that is, with a reasonable degree of payback given the conditions of the project. One of the site managers emphasized the multiple skills demanded in his work:

> You need to have a number of skills. I do not think it is enough to have just one but you need a few of them. You need to be able to listen to people, speak to them, make the clients confident, and when dealing with authorities, then you cannot do as you like; you need to be a bit agile and not just move on because then you don't have that client any more. Customer relations are part of my work.
>
> (Site manager 7)

The site manager continued:

> When you start here as a site manager, there is a great deal of demand on you. If you are a site management at ConCo, then you really need skills in many, many areas: work occupation laws, environmental law, emission restrictions, you name it. You talk to all sorts of authorities and they are chasing you ... These guys [the site managers] need to be updated.
>
> (Site manager 7)

Besides the need for know-how in rock construction and project management, the site managers also emphasized the value of knowing all the 'tricks of the trade' that, as one site manager (no. 7) put it, 'can save you immense amounts of time and money'. For instance, when conducting drainage work in a tunnel, there are many small things that save time and reduce costs. In addition to the knowledge content of the work, the work was also perceived as being to some extent unpredictable. One of the site managers argued it was complicated to predict what would happen in some projects:

> You can never formulate a description that is complete – maybe in some jobs where you do something easy but if you are doing a few more complex things then you cannot write it all. Many conditions are not fully known when you start a work ... it is enough to get into the ground

when we are rock reinforcing, you just don't know what it looks like. They [the client] may have done a few tests in two, three places but that don't count for much.

(Site manager 8)

A third source of uncertainty was that there were, as in many construction companies and other firms, being 'projectified', a lack of adequate evaluations of how previous projects were carried out and whether they met their financial goals. The Chief Executive Officer (CEO) admitted that ConCo had not been 'very successful' in implementing procedures for project evaluations, notwithstanding a general belief that such procedures would be helpful in the long term:

It goes something like this: 'Didn't we have this kind of project like three years ago? Didn't you participate in that project?. Could you please go and get the binders so we can see how we did it back than, and then we can use the same calculation' ... There is this basis of knowledge but it is not systematic.

(CEO)

One of the site managers shared this view and claimed it worked in this rather informal manner:

This is how it works: You walk over to a colleague and ask: 'Your white book [the formal documentation of the project] for project X, can I borrow it to see what happened?' If someone should take care of these things [scanning of documents and making them available in data bases], then we have to hire an additional person and I wouldn't say that is necessary.

(Site manager 3)

Taken together, the complexity of the work procedures, the unpredictability of the construction work, and the lack of systematic project evaluation procedures produced a work situation where site managers were trained to conceive of their own role as being 'problem-solvers' capable of handling emerging situations in the best possible manner. Rather than assuming that the project would run in accordance with pre-specified rules and objectives in a linear and rational manner, site managers knew that they were expected to handle a number of unanticipated events during every workday. In order to handle all such emerging concerns, site managers essentially relied on the expertise and experience of their colleagues both internal and external to the firm. The continuous flow of information between the site-workers therefore demanded a certain degree of trust and confidence. One of the site managers praised the open attitude in the company: 'We are very open towards one another. We are tolerant. All failures may be discussed'. The conversations where the site managers exchanged insights and know-how were not deliberate or planned events but normally took place in very informal settings. Several of the co-workers emphasized the coffee table conversations as an important source of knowledge-sharing:

> We are a small company. We sit here and we have coffee together and if there is anyone facing a problem then we need to discuss it with one another. People come here to ask 'have you done this before?'
> (Site manager 5)

Another site manager referred to the coffee breaks in similar terms:

> A: It is quite important to have a close dialogue with one another to avoid these problems.
> Q: Where do you have this dialogue?
> A: By the coffee table.
> (Site manager 6)

While the coffee table meetings were valued by most of the interviewees, they were still not official arenas for knowledge-sharing, that is, much know-how that may be of value for the site managers may not be surfaced and articulated in a more systematic manner. As site manager 6 pointed out, his colleagues are 'not refusing to share their insights but they are not really sharing without being asked about it'. The informal knowledge-sharing routines centred on the joint coffee break were an important arena in the site-manager group. Still, some of the more critical site managers emphasized that there may be a value in instituting more formal procedures and evaluations of 'lessons learned' as an integral part of each project: 'The whole of ConCo is based on knowledge but there is actually no real forum where you collect all new knowledge', site manager 9 noted. At present, ConCo could be perceived more or less as an 'oral culture' (see Chapter 2) where verbal communication is the principal media for conveying know-how. Oral communication is very effective, but it is also of necessity local and context-bound. 'A lot is never documented anywhere', site manager 14 remarked. As site manager 13 pointed out, 'the problem occurs when people quit and they have not documented their work anywhere'. The standard argument for not using more standardized evaluation procedures was articulated along two rationales. First, site managers were already all too burdened with various forms of 'paperwork', essentially not the type of work they favoured or thought of as being rewarding, and therefore they were not willing to add extra work of this kind: 'They [project evaluations] might have a value to preserve but it is quite heavy to sit and do all the work so actually it's not really done', site manager 8 said. The other argument was that each project was more or less unique and reappeared only after long time periods. The Stockholm branch division manager represented this position:

> Some things reappear like every tenth to fifteenth year. If there is no one being experienced from this kind of work, then we just don't know what to do ... Some kind of photo documentation accompanied by comments [would be helpful]. I run like 60 projects every year.
>
> (Division manager)

Since site managers did not appreciate additional paperwork and individual assignments may not reappear in some 10 to 15 years' time anyway, the co-workers at ConCo preferred to dedicate time to trace individual experts in the industry when they really had to. When being asked if the expertise is lost every time an idiosyncratic project is terminated, the division manager responded that such was not necessarily the case, because they were 'good at tracing those who worked on the project 15 years ago'. One of the site managers specializing in rock reinforcement work told a supposedly representative story of how they managed to handle a major setback in an advanced tunnel construction:

> We were injecting poles down to the rock to avoid the whole tunnel floating up. We drilled through the concrete and onward down to the rock to attach a wire to the bottom. When we were 50 meters down in the rock, all of a sudden we lost the hammer [a tool used in rock reinforcement work] and we had to get it back up to the surface. There was no alternative because we could not move the hole. But what do you do when something weighing 200 kilos is stuck 50 meters below you in the rock? That was a good test. It took three days ... We had to call everyone and ask 'how do you do this?'
>
> (Site manager 5)

An openness to the unexpected and a great deal of social capital manifested in and operationalized in a network of construction work experts are what helped in solving the precarious situation when the tool got stuck deep down underground. This story is representative of how site managers are capable of dealing with situations they have not previously encountered and underlines the importance of the ability to share know-how in the broader community of site managers or construction workers more generally. The unwillingness to formalize and 'mediatize' (i.e. to store project evaluations in, for example, databases) know-how among site managers is representative of what Strauss, Schatzman, Bucher *et al.* (1964) call a *professional ideology*; the legitimate and professional site manager is not trained to predict and prevent any possible event because it is futile to believe that is possible to do. Instead, it is the ability to handle upcoming situations and to make use of collective resources of know-how and expertise that defines the skilled site manager. Using Hoopes and Postrel's (1999) helpful term 'glitch' (a Yiddish term meaning 'slippery area'), denoting 'an unsatisfactory result on a multi-agent project that is directly caused or allowed by a lack of interfunctional or interspeciality knowledge about problem constraints' (Hoopes and Postrel, 1999: 843), site managers are the experts in sorting out and handling all the problems and concerns – the glitches – that emerge in between the various professions and occupational groups. Hoopes and Postrel use the term 'glitch' in their study of product development work, but the term is equally applicable in a construction industry setting. Site managers negotiate relations between conflicting interests and participate in a significant number of decision-making activities, in many cases on the basis of limited information and under substantial uncertainty. Believing that one can decode and structure this capacity to handle glitches and ambiguous situations in documents and templates is largely at odds with the professional ideology of the site manager (see e.g. Rooke, Seymour and Fellows, 2004). In addition, site managers participate in an oral culture wherein the telephone is the principal medium for communication and for gaining access to the broader network of site managers. 'I speak in my cell phone most of the time ... It is red hot, right', site manager 8 remarked. Such an oral culture is ultimately embedded in the collective social capital accumulated and mobilized in the community of site managers. Social capital is in this view the underlying resource that enables the day-to-day work of site managers.

The influence of social capital in rock construction work

Knowledge-sharing is in many respects a paradoxical social procedure; on the one hand, knowledge is in essence what is social and what floats around in social communities; on the other hand, numerous studies of knowledge-intensive firms show that knowledge is in various ways 'inert' and 'sticky' (Szulanski, 1996; Von Hippel, 1998), not really being distributed and shared between individuals, departments, projects or collaborating firms (in, for example, supply-chains or in joint ventures). Following the empirical study of Subramaniam and Youndt (2005), it is here assumed that social capital plays a pivotal role in mediating human capital (in Subramaniam and Youndt's, 2005, sense of the term), and the organizational capital. Without investing in social capital, the stocks of human and organizational capital are not fully exploited and used. Social capital is a term denoting a number of shared and collective emotional, cognitive, and communicative skills and resources that help to circulate and develop existing know-how in a firm. In ConCo, the knowledge mobilized and used was regarded as being contingent, context-bound and embedded in practical experience. In addition, the diversity of projects – one specific type of project may come back every ten to fifteen years, one of the company representatives suggested – and the unwillingness among the site managers to engage in formal evaluations of terminated projects constituted a 'knowledge system' that demanded a close collaboration between the community of site managers. The main form for sharing insight, experience and know-how was therefore oral communication, either in direct face-to-face discussions around the coffee table, or over the telephone. In either case, know-how was conveyed verbally and in a practically oriented manner. For proponents of advanced computer-mediated knowledge-sharing practices, this reliance on the verbal interaction and storytelling may sound outmoded or even archaic. Contrary to such dismissive views, the oral and storytelling culture of site managers, the central social practice enabling knowledge-sharing, may be regarded as the most effective way to convey learning and insights given the strong influence of idiosyncrasies and context-bound conditions in the rock construction projects managed by the site managers. The findings of the study thus support a more affirmative view of 'everyday talk' in organizations presented by, for example, Currie and Brown (2003), Patriotta (2003), Bartel and Garud (2003), Orr (1996), Donnellon (1996) and Boden (1994). However, rather than talking about the site managers' conversations and storytelling in strict narrative terms (see e.g. Czarniawska, 2004), the concept of social capital includes both the collective capacity to tell stories, but also other non-verbal competencies and joint accomplishments such as trust, shared professional ideologies and norms. Knowledge-sharing in the ConCo is thus embedded not only in verbal communication but in a multiplicity of social practices and beliefs that structure both the construction industry, the focal firm and its partnering companies (e.g. suppliers, sub-contractors) but also the professional and occupational groups. Expressed differently, social capital is the 'infrastructure' (Bowker and Star, 1999) of ConCo and the construction industry alike in which ConCo operates. If one wants fruitfully to examine and understand how construction firms develop, share and exploit their intellectual capital, then the concept of social capital needs to be brought into the analysis.

Summary and concluding remarks

Many construction companies struggle how to do more with less resources and one of the most significant domains when substantial effectiveness increases may be gained is in knowledge sharing. This study suggests that social capital plays a decisive role as the infrastructure for all knowledge sharing activities. In ConCo, knowledge-sharing is essentially verbal, either based on direct face-to-face communication or over the telephone, and there is little interest in developing more formal models or procedures for knowledge-sharing among the ConCo employees. In other words, the site managers were expected to act as a community of practice (Wenger, 2000; Wenger, McDermott, Snyder, 2002; Lindkvist, 2005) capable of helping one another to sort out things and solve problems as they occurred *en route*. This study contributes to a perspective on knowledge management, and more specifically to the literature on knowledge-sharing, though emphasizing the value of an open, trustful and collaborative attitude among professional or occupational groups relying on their ability to exploit know-how and previous experiences in their work. Site managers in construction companies is one such occupational group, taking pride in their ability to run complex projects without any major disturbances and to handle all sorts of challenges presenting in their day-to-day work. For the community of site managers, a broad network of colleagues and associates is of central importance, and what is also accumulated over years of experience.

Aesthetics in rock construction work

Introduction

In a recent move to broaden the basis for organizational theorizing, the concepts of the aesthetic and 'aesthetic knowledge' have been invoked as complementary to the conventional proposition-based and logical modes of thinking privileged in the Western tradition. Böhme (2003) even speaks of the emergence of what he calls an 'aesthetic economy', a concept taking seriously the 'fact' that 'aestheticization represents an important factor in the economy of advanced capitalist society' (Böhme, 2003: 72). Since capitalist production has attained a particular stage of development and no longer serves to satisfy individuals' 'material requirements', it must instead 'turn to their desires'. As a consequence, Böhme (2003: 72) argues, the economy becomes an 'aesthetic economy'. Among various changes in the aesthetic economy, labour is characterized by an increase in aesthetic content. 'Aesthetic labour', Böhme (2003: 72) says, 'designates the totality of those activities which aim to give an appearance to things and people, cities and landscapes, to endow them with an aura, to lend them an atmosphere, or to generate an atmosphere to ensembles'. In late modernity, capitalist production is in essence drawing on aesthetic resources.

In the literature on organizational aesthetics, the ability to think and perceive in aesthetic terms is rendered not only a legitimate but also a highly valuable skill and capability in organizational settings (Hancock, 2005; Hancock and Tyler, 2007; Mack, 2007; Warren, 2008). For instance, Taylor and Hansen (2005: 1213) argue: 'Aesthetic knowledge offers fresh insight and awareness and while it may not be possible to put into words, it enables us to see in a new way'. Speaking of aesthetic knowledge in architecture work, Ewenstein and Whyte (2007a) emphasize that this human faculty derives not so much from linguistic resources and the use of language, but from a 'conceptual space', including the interaction between perception, materiality, emotionality and tactile, audible, and olfactory skills and capabilities constituting what they call 'aesthetic reflexivity:

> [A]esthetic knowledge and knowing goes beyond word: including both symbolic and experiential forms ... aesthetic reflexivity opens up a conceptual space in which to explain the mechanisms through which knowledge begins to emerge in interactions with materials and other actors.
>
> (Ewenstein and Whyte, 2007a: 705)

In addition, an emergent body of literature addresses aesthetics not only as a source for theorizing but claims that the production of aesthetic work such as art and performing arts are in themselves the outcome from or an integral component of organized activities. Adler (2006) discusses the value of aesthetic training in leadership work; Guillet de Monthoux (2004) examines what he refers to as the 'art firm', organizations operating in various aesthetic domains; Gibson (2002) shows how art programmes were used during the depression in the 1930s in America; Van Delinder (2005) points out the co-evolution of Taylorism and the modern Russian ballet of George Balanchine and Igor Stravinsky in the 1910s. The relationship between aesthetics and organization is multiple and variegated. The concept of aesthetic knowledge, and aesthetics in more general terms, offers a new set of concepts and a vocabulary for organization studies and serves to integrate a number of scattered theoretical frameworks, including emotional management, design studies, arts management, and theories of embodiment or perception in organizational settings.

The concept of aesthetics

The concept of aesthetics has been part of the continental philosophy tradition since Alexander Baumgarten, Immanuel Kant's tutor, formulated a philosophy of the aesthetic. However, as Eagleton (1990) points out, in Baumgarten's work, the term 'aesthetics' does not in the first place refer to art *per se* but to the whole region of human perception and sensation in contrast to conceptual thought. Eagleton (1990) argues:

> The distinction which the term 'aesthetic' initially enforces in the mid-eighteenth century is not between 'art' and 'life', but between the material and the immaterial: between things and thoughts, sensations and ideas, that which is bound up with our creaturely life as opposed to that which conducts some shadowy existence in the recesses of the mind.
>
> (Eagleton, 1990: 13)

Today, the term aesthetics has taken on a more general meaning as 'the philosophy of art and beauty' (Shusterman, 2006: 237). The various historical and contemporary uses of the term aesthetics are still making it a fundamentally multiple term, Shusterman (2006) argues:

> The concept of aesthetics remains deeply ambiguous, complex and essentially contested. This is partly because the notions of art and beauty themselves involve such ambiguity, complexity and contestation, but also because the notion of aesthetics has an especially complicated, heterogeneous, conflicted and disordered genealogy.
>
> (Shusterman, 2006: 237)

In the German Romantic tradition, aesthetics played a central role as the human skill or capacity to apprehend and perceive the beautiful. The best-known defence of the value of 'aesthetic training and education' from this period is Friedrich Schiller's *Letters on Aesthetic Education* (1795) in which he declared that aesthetic training is the gateway to rational thinking and reason: 'In a word, there is no other way to make the sensuous man rational than by first making him aesthetic', Schiller (1795/2004: 108) claimed. In the European tradition, Schiller's *Letters on Aesthetic Education* has been enormously influential in terms of providing a strong case for the need for culture and education in the domain of aesthetics. However, in the European canon, there is also a strong line of demarcation between, on the one hand, sensuous, aesthetic experiences (*aistheta*) and the logically grounded and rational modes of thinking of the sciences (*noeta*) (Luhmann, 2000a: 15). More recently, the boundaries between aesthetics and science have been loosened and several studies of scientific work emphasize the aesthetic components in all scientific activities. For instance, Hallyn (1990: 159) argues that Copernicus's theories of the universe satisfied the aesthetic requirements that are characteristic of the classical Renaissance: 'symmetry' and the perfection of the circle. Aesthetic criteria thus penetrate the domain of the sciences. Furthermore, Jordanova (1989) claims that academic medicine and especially surgery was strongly influenced by predominant aesthetic ideals during the nineteenth century:

> The aesthetic dimensions of science and medicine are beginning to be paid the more serious attention they deserve, not to display them as cultural ornaments but to demonstrate that aesthetics is constitutive of knowledge. We can see, for example, that 'realism', used as an aesthetic rather than a philosophical term, has been important within science and medicine by defining modes of illustration, and also, conversely, that scientific and medical ideas were central to realist artistic and literary practices, especially in the nineteenth century.
>
> (Jordanova, 1989: 6)

Even though scientific ideologies impose a value-free or detached position *vis-à-vis* the object of study, the relationship between aesthetic and science remains complicated: 'Art and science are ... at once distinct and connected. They coincide in one aspect, the aesthetic aspect. Every scientific work is at the same time a work of art', Croce (1992: 27) summarizes his argument. Applying Croce's (1992) argument to technology, Kasson (1976) argues that the defence and praise of modern machinery in nineteenth-century America included appeal to 'aesthetic criteria'; even 'habitual and highly trained observers frequently expressed delight in the spectacle of productive technology', Kasson (1976: 140) notes. For the public at large, distrusting the fine arts, the modern functioning engine represented a product:

> capable of eliciting intense aesthetic enjoyment, that was the instrument not of decadence and tyranny but of a progressive, republican nation, the consequence not of idleness and expense, but of industry and ingenuity, in short, the reflection not of Europe and the past but of America and the future.
>
> (Kasson, 1976: 146)

As Kasson (1976) notes, throughout the nineteenth century, the very word 'arts' encompassed skilled craft generally, including invention, and in the medieval times, Le Goff (1993: 62) remarks, all intellectual work was included in the domain of *ars* and *techne*. The concepts of 'artist' and 'artisan' are testifying to the shared etymology. However, over the course of the century the increased uses of qualifying adjectives 'fine' and 'useful' broadened the gap between the arts and the crafts (Kasson, 1976: 146). Kasson (1976) concludes his argument:

> Many of the same aesthetic values Americans demanded in the fine arts they sought also in technology. As a result, form followed not only function but also fashion and symbolic expression to a surprising degree in nineteenth-century American industrial design; so too, to an even greater extent did observers' interpretations of mechanical form.
>
> (Kasson, 1976: 154)

Technology was closely associated with concepts such as beauty and the aesthetic; it embodied not only human reason and rational thinking but also appealed to other human faculties.

Aesthetics and organization

In organization theory, there has been a more articulated interest in the concept of aesthetics during the last few years, beginning in the 1990s (Strati, 1999; Linstead and Höpfl, 2000; Taylor and Hansen, 2005). In this discourse, aesthetics is serving as a portmanteau term, including a series of perceptual and emotional human skills and capabilities complementing the conventional rational and logical reasoning predominating in organizations and corporations. What has been called 'aesthetic knowledge' (Hancock, 2005; Ewenstein and Whyte, 2007a) is thus what operates on the basis of embodiment, intuition and emotionality, and what even escapes pre-existing vocabularies and regimes of representation. For instance, 'aesthetic knowledge is embodied, it comes from practitioners understanding the look, feel, smell, taste and sound of things in organizational life', Ewenstein and Whyte (2007a: 689) write. Taylor and Hansen (2005: 1213) speak of aesthetic knowledge in similar terms: 'Intellectual knowledge is driven by a desire for clarity, objective truth and usually instrumental goals. On the other hand, aesthetic knowledge is driven by the desire for the subjective, personal truth usually for its own sake'. Many contributors to the discourse on aesthetic knowledge underline the importance of mobilizing and recognizing the five senses and not just relying on cognitive faculties and rational thinking. Marotto, Roos and Victor (2007: 410) claim:

> Beyond rational understanding of the task at hand, it is important that group members perceive and experience each other's work aesthetically. This calls for a deeper appreciation of how and what we perceive through our five senses as well as the subtleties and ambiguities of interpersonal communication and cooperation.
>
> (Marotto, Roos and Victor, 2007: 410)

Empirical studies of organizational activities and undertakings from an aesthetic knowledge perspective include ethnographies of architecture work (Ewenstein and Whyte, 2007a), studies of a symphonic orchestra (Marotto, Roos and Victor, 2007) or the semiotic analysis of recruitment material from a major accounting firm (Hancock, 2005). Some studies find aesthetic elements in unexpected places such as in computer programming (Piñeiro, 2007), financial services (Guve, 2007) or restaurant chef work (Fine, 1996). Piñeiro (2007) suggests that even though computer programming is commonly associated with the formulation of strictly logical instructions (see e.g. Ullman, 1997) there is in fact space for aesthetic concerns:

> Even in the exact and demanding world of computer programming, there is a place for questions of an aesthetic nature. Regardless of the restrictions forced upon human activity, if there is as much as an ounce of creative work involved or permitted, aesthetic concerns will thrive.
>
> (Piñeiro, 2007: 105)

For computer programmers, there are certainly 'beautiful programs' and there are 'ugly programs'; the perceived beauty of certain programs derive from a variety of ideologies, opinions and beliefs regarding what constitutes effective and skilful programming. In general, however, beautiful programming is the economic use of instructions and skilful solving of problems occurring *en route*. For Piñeiro (2007), the aesthetic of programming is representative of the need for recognition and status in the computer programmer community:

> Why would programmers want to write beautiful code? The immediate answer is: programmers do not perceive code as only a virtual machine that does things but also as their creation. Their relationship to code is more that of creator than of technician to machine. Their code then speaks of them: not only of their skills but also their personal preferences in things like coding style, programming language and designing strategies.
>
> (Piñeiro, 2007: 119)

Even in the domains of society where logical and rational thinking predominates, there are pockets of aesthetic concerns guiding social actions and human beliefs. The various forms of aesthetic knowledge are therefore not a marginal or peripheral resource in organizations, but play an active role in many activities and operations. In the following, the work of rock construction workers will be examined in terms of their aesthetic knowledge, that is, the ability to perceive particular qualities in the work of the skilled construction worker, not only in terms of functionality but also in terms of its aesthetic features.

The beauty of sprayed concrete, rock injection and bolt reinforcement

In ConCo, it was generally held that the spray-concrete industry had changed considerably since the 1950s and 1960s in terms of new technology and new forms of concrete. One of the interviewees emphasized the radically higher productivity in today's industry, saying that 'two workers spray what ten used to do. And is not that heavy work any more, holding a hose spraying concrete … Many things have changed' (Maintenance and warehouse worker). In addition, it was generally claimed among the site managers that construction workers employing the technology were the very centre of the operations; without skilled and specialized construction workers, quite simply, no ConCo. One of the newly hired site managers with a Masters degree in civil engineering said:

> You must respect the construction workers. If you don't know them, then you know no one, and if you do not get along very well with them, then you are not getting along well with anybody – that's what it's like.
>
> (Site manager, ConCo)

The skilled construction workers were highly valued in ConCo and in the industry because rock construction and sprayed concrete is a very specialized field and it takes a long time to train workers to use the equipment and master the work fully. One of the site managers stressed this predicament for the company:

> We are an expert company; not too many are capable of doing what we do. When we are busy, then we cannot just call the employment agency and say 'now we need a drainage guy' or 'we would like to hire a robot operator' because we know there is no such thing. We *always, always* have to train our staff, all the time. And when they are skilled after a few years, then the major corporations in the field pick them out and offer them better pay. I just lost my best drainage guy to Norway.
>
> (Site manager 9)

The CEO also deplored the lack of formal training, saying that 'there is no formal training ... dedicated to shotcreting because the market is simply too small'. Instead, to work with rock construction and more specifically with sprayed concrete was a tradition in some families. The CEO pointed out the value of such tradition for the firm:

> I believe our competitive advantage is that we are very flexible, and have, especially in Stockholm, competent construction workers ... we employ like three, four guys working here in the third generation in ConCo. That means that we have a high degree of competence and that is something the clients appreciate.
>
> (CEO)

Learning to use the robots and the other advanced equipment was generally regarded a highly specialized skill, not really accessible for anyone to learn and master fully. When discussing the problems associated with recruiting and maintaining a skilled cadre of construction workers specializing in rock construction and sprayed concrete, the site managers and other company representatives draw on what has been called 'aesthetic knowledge'. While otherwise being strongly shaped by professional ideologies stressing the value of 'hands-on skills' and rational thinking, the site managers conceived of the work of the best spray-concrete robot operators in terms that are best described in an aesthetic vocabulary. First, the CEO emphasized the tacit knowledge and the 'sleight of hand' of the best construction workers:

> The key skill is that you are capable of doing the work ... What we do at ConCo, injecting tunnels with concrete, then you need to have a certain sleight of hand and decide when the pressure is good and ... you need to be skilled to manage the work. You and I can plant a lawn or plant tulips – we won't do it as fast as a professional, however – but we cannot inject with concrete. It is too complicated, quite simply.
>
> (CEO)

Several site managers shared this view and emphasized that there was a strong influence of a tacit component in the work:

> Some operators are not capable of learning certain things. They do not have the right touch ... Among the best one to hire are farmer boys; they have grown up with various machines and have learned to tinker with the machinery and they have a certain feeling for it. Then there are others who have been here for like ten years and still don't know a thing because they lack that skill.
>
> Q: So they lack a real interest?
>
> A: Well, in fact, they can be *very* interested. But they do for instance lack the ability to listen to the machine and how it penetrates the ground.
>
> (Site manager 5)

'Either you are capable of operating or you are not. Some operators never make it', another site manager (14) claimed quite flatly. One site manager argued that some operators could do a great job on the work site but did not take into account how to avoid dirtying the whole area while working, leaving the place in a mess; on the other hand, some of the construction workers considered all conceivable details in the work but yet failed to accomplish a satisfying result. Only a few of the construction workers could handle the entire spray-concrete process, both 'setting the scene' for the work and providing an appealing end-result:

> Manual spray concreting and such things, you can never learn to master but you need to know it from the beginning. Either you know how to do it or you don't; you'll never learn unless you know it from the outset ... I've seen many lads come and go here. Some try and try but it still doesn't work and for them it is just painful to continue. To see the whole spectra, a final product [is problematic]; they may perform a certain component of the job brilliantly, but then they don't give a damn thing about anything else ... And that happens over and over: 'Oops, I didn't think about...' ... We have hired people who have been on the other end of the continuum, who are very detailed about everything and yet as soon as they start everything just fails and then it costs a substantial amount of money to get it right.
>
> (Site manager 9)

Some of the site managers emphasized various tactile, visual and audible skills for being able to accomplish a qualified work. One of the site managers explained:

> To operate the machinery is not that difficult but to do it good is damn hard. Then you need substantial experience ... Learning how to change

the thickness of the concrete, if to speed up or slow down or whether to lower the pressure when injecting.

(Site manager 11)

Another interviewee, the maintenance and warehouse worker responsible for repairing and maintaining the machines, emphasized the visual and audible skills when operating the spray-concrete robots:

Quite often, you hear various sounds ... You may see the movements of the machine ... You notice that it [the machine] doesn't do too well, the spray concrete just bubbles and it hisses by the mouth-piece, and then something's wrong ... You can listen to the pump-beats whether machine works as it should.

(Maintenance worker)

'Knowing the machinery' included the ability to diagnose any abnormal or unexpected events prior to a major breakdown. Since the rock construction work operated on short project times, in many cases just a few days, a breakdown of the equipment may mean the loss of the entire profit in a single construction project. The ability to take care of the machinery and equipment was therefore of key importance. To use what here has been called the 'aesthetic knowledge' is therefore important for the day-to-day operations. In the site managers' and the other ConCo representatives' articulation of important skills and competencies, a strong but implicit line of demarcation between standard operation procedures and the craft of the work is emphasized. The standard operation procedures may appear fairly standardized and repetitive, but the ability to perform them skilfully is part of a craft that has, at best, to be trained and, at worst, to be a talent you are endowed with. Seen in this perspective, it is not too far-fetched to argue that there is an 'aesthetic of rock construction' that is recognized by the site managers. However, when speaking of aesthetic knowledge in this community, it denotes something quite different from that in, say, a community of architects. Among the rock-construction engineers, aesthetic does not mean to be capable of demonstrating a familiarity with theories of design or architecture or an awareness of the differences between, say, the works of famous architects such as Le Corbusier or Frank Lloyd Wright. Instead it signifies a more mundane and down-to-earth capacity to take care of operations in a skilled manner; to set up and carry out an operation effectively and to perform professional work satisfying or even exceeding – a point emphasized by the CEO – the expectations of the clients. In ConCo, aesthetic knowledge represents an integral component of qualified workmanship, that is, the ability to take on an assignment and carry out all the operations demanded. Rather than using the etic term 'aesthetic knowledge', the ConCo co-workers used terms like 'feeling' to signify the integration of a range of human faculties in the day-to-day work.

For instance, one site manager used this term: 'Rock injecting is a very specialized niche in the market. There, you need to have a very good sense of *feeling* [English in the original] to understand what happens when you are injecting' (Site manager 14). Such 'feelings' are not marginal or peripheral to the work of rock construction workers, but appear to play a central role in carrying out day-to-day work.

The work of rock construction workers shares many interesting features with other professions and occupational groups engaging cognitive, perceptual and embodied skills. For instance, Prentice (2005) examines the use of virtual reality simulations in surgery training, and she argues that experienced surgeons develop what she calls 'somato-conceptual intelligence', the seamless integration of cognitive capacities and tactile skills: 'With years of practice, surgeons learn to use tools as extensions of their bodies. Technique becomes fully embodied and, therefore, largely unconscious, when all proceeds smoothly' (Prentice, 2005: 856). The surgeon knows all parts of the human body of relevance for his or her expertise, but so do their hands (see also Merleau-Ponty, 1962; Sudnow, 1978). Surgery is here conceived of not only as an intellectual or cognitive activity, but is equally what is embodied and somatic; the connections between brain, eye and hand are central to skilful surgery work. Without drawing conclusions that are too far-fetched, rock construction workers, especially those specializing in sprayed concrete, develop a similar form of somato-conceptual intelligence; they draw on sense-impression, previous experience and embodied performances in their work. They also have to develop certain sensitivity when using the technology and materials. The concrete has specific qualities, such as viscosity, density and homogeneity, and the machinery has to work smoothly to provide the best result. Just like the surgeon is capable of interacting effectively with the patient's body, the rock-construction worker creates a relationship with the rock tunnel wall he is working on. Hence the importance of analytical constructs such as 'aesthetic knowledge'.

Aesthetic knowledge in non-aesthetic work

In the general recognition that the basis of human knowledge does not necessarily derive from cognitive and rational thinking but is grounded in visual, tactile, audible and olfactory perception in organization theory, the concept of aesthetic knowledge is useful to denote a variety of such residual categories. The concept of aesthetic knowledge thus opens up a new perspective on organizational knowledge and broadens the basis for the analysis. For instance, aesthetic skills and competencies are no longer solely located in specific industries and occupational groups in the domain of design and arts management, but are instead what inform, shape and affect a variety of professions, including financial analysts, computer programmers and restaurant chefs. The study reported here contributes to this literature through emphasizing the aesthetic components in rock construction work.

Rock construction work arguably shares very few characteristics with the conventional professions and industries commonly associated with the concept of the aesthetics (e.g. design, the arts, interior decorating and so forth). Instead, rock construction is in essence what may be regarded as non-aesthetic or even anti-aesthetic; it is a dirty work, operated by working-class male construction workers, taking place under unfavourable conditions in tunnels, cellars and other subterranean domains. In addition, it is work that is rarely, if ever, accounted for in aesthetic terms. The work of rock construction workers is 'infrastructural' rather than ornamental; this implies that what rock construction workers accomplish is by definition largely taken for granted. Similar to the electricity companies providing heat and lighting, or road and railway planners and maintenance workers, rock construction workers' accomplishments are always in the public mindset, already in place, ready to use and take advantage of. However, notwithstanding the relative ignorance on part of the broader community, rock construction workers take pride in their work and achieve credibility on the basis of their ability to perform their work in an aesthetically qualified manner; the ability to structure and perform the work in an effective way and to accomplish an aesthetically appealing job is what gives prestige and privileges in the construction industry.

At the same time as the concept of aesthetic knowledge is helpful in explaining the qualities of rock construction work that no other vocabularies can fully capture, there may be other terms denoting such skills. It is noteworthy that the concept aesthetic knowledge is by no means an *emic* concept, a term used by construction workers themselves, but an *etic* concept, part of the outsider analyst's vocabulary. Rock engineers would talk about 'doing a nice job', 'meeting the standards' or, most likely, 'satisfying the client's expectations'. In their view, aesthetic knowledge is probably all too closely associated with the fine arts and middle-class culture to be fully credible to use. When site managers in construction work talk about the skills of their best employees and co-workers, they favour expressions like 'knowing the job', 'he can really get his way through' and 'being capable of handling the situation'. Such expressions and ready-made articulations are, however, not devoid of or in conflict with aesthetic norms and qualities, but they are rather innate to the very term *skill* (Attewell, 1990). Even though the construction industry is one of the few largest industries capable of maintaining a craftwork tradition, the distinction between *artist* and *artisan* is clearly demarcated after decades of divergence between the arts and the trades (Kasson, 1976: 146). Bringing aesthetic concern back into the analysis is, therefore, informing to broaden the analytical repertoire used in studies of workplace learning and knowledge management. Rather than assuming that there are few or no aesthetic components in work and occupations not commonly associated with the aesthetic, the concept of aesthetic knowledge underlines that aesthetic standards and aesthetic thinking plays a relatively significant role in many occupations and professions.

Summary and conclusion

The study of ConCo suggests that rock construction workers demonstrate varying capacities for accomplishing a satisfying result in their work. The term 'aesthetic knowledge' is here what may enable an understanding of the skills more successful rock construction workers include in their professional repertoire. The ability to see, hear, smell and touch thus complements the more analytical and routine-based thinking that constitutes much construction work. This study thus contributes to an emerging literature on aesthetic knowledge in what is generally conceived as being non- or even anti-aesthetic work. Similar to computer programmers or financial analysts, rock construction workers develop and use certain aesthetic skills that help them accomplish their day-to-day assignments without generally being credited for these capabilities. Broadening the basis for the analysis of aesthetic work to new and previously little considered domains may provide new insights and learning on how aesthetics are a central feature of many occupations and professions.

6 Institutionalizing knowledge in construction work organizations
Theoretical and practical implications

Introduction

In this final chapter, some of the implications from the four studies of the construction industry will be examined. In this discussion, the concepts of institutionalization and, more specifically, the institutionalization of knowledge will be used to show how knowledge is not just emerging *ex nihilo*, but is instead carefully embedded in pre-existing routines, practices, roles and ideologies already in place in a social field. In the first chapter of the book, a sociological view of knowledge as what is always based on practice, which in turn is strongly determined by institutions, such as professional and occupational ideologies, and organizational routines and standard operation procedures, was discussed. Following this line of reasoning, this chapter suggests that the institutionalization of knowledge is by no means an uncomplicated or predetermined process, but is instead a process affected by the interests and power of the social groups and communities involved in the institutionalization process. The institutionalization of knowledge is therefore characterized by a combination of conflict and dissent, and consensus and joint agreements. After this more theoretical discussion, some practical and managerial implications are addressed.

The institutionalization of knowledge

In the knowledge-management discourse, there are a great variety of ontological, epistemological and methodological views of the social resource named knowledge. In the following, rather than assuming that knowledge is always already in place, ready to apply to cases, knowledge will be examined as what resides, at least partially, in various institutions in society. As pointed out in the first chapter, knowledge is closely bound up with professions, practices and social action; conceiving of knowledge as a *flow* rather than a fixed *stock* is thus to underline the social and collective nature of all knowledge. In this final and more analytically oriented chapter, knowledge will be examined as what is institutionalized in the routines, rules, standard operation procedures and practices enacted in organizations. Following Gerardo Patriotta and Giovan Francesco Lanzara (2007), all knowledge is stabilized and rendered legitimate and thus uncontested through processes of institutionalization. In the construction industry, a great number of standard operation procedures and scripts for action prevail, embodied in the everyday work of site managers, construction workers, designers and architects. To learn and to decode and encode such practices is one of the principal challenges but also sources of learning for any newcomer to the industry. Expressed differently, any legitimate and generally accepted agent in the construction industry – or any other industry, for that matter – is familiar with the institutional setting in which he or she participates; the instituted milieu is what is experienced as what is, to some extent, taken for granted and therefore unproblematic for the agent affiliated with this particular industry.

The field of institutional theory in organization theory is one of the oldest and arguably most prestigious theoretical fields in the management studies literature. A range of academic disciplines, including sociology, political science, economics, anthropology and psychology, has contributed to the field. Beginning with the works of Talcott Parsons (1934/1990) and later on the work of Philip Selznick (1949), in the 1970s and subsequently the concept of *new institutionalism* (Selznick, 1996; Hasselbladh and Kallinikos, 2000) was coined and advanced by seminal papers by Meyer and Rowan (1977), DiMaggio and Powell (1983) (for an overview, see Zucker, 1987; DiMaggio and Powell, 1991; Scott, 1995; Tolbert and Zucker, 1996). The line of demarcation between 'old' and 'new' institutional theory is a rather porous one, but Scott (2004) suggests that while:

> earlier theorists such as Selznick (1949) and Parsons (1990) stressed the regulative and normative aspects of institutionalized systems ... later neoinstitutionalists recognized these as significant factors, but they also called attention to the role of symbolic elements – schemas, typifications, and scripts that perform an important, independent role in shaping organization structure and behavior.
>
> (Scott, 2004: 7)

Recently, institutional theory and the more specific concept of institution have been invoked in a range of studies of innovation (Leblebici, Salancik, Copay and King, 1991; Nooteboom, 2000; Hargadon and Douglas, 2001; Whitley, 2000) and organizational practices (Guler, Guillén and Macpherson, 2002). In the next section, the concept of institution and some accompanying concepts will be examined in some detail. Thereafter, the function of institutions for organizational structure and managerial practice will be discussed, and finally, the institutionalization of knowledge is covered.

Defining 'institution'

The first thing to define in the analysis is the concept of 'institution'. Institutional theory was one of the dominant theoretical frameworks in the 1980s and 1990s in political science, management studies, sociology and a few other social science disciplines (Scott, 1995), effectively propelled by the publication of DiMaggio and Powell's (1991) edited volume *The New Institutionalism in Organizational Analysis*, suggesting a 'new institutionalism' framework of analysis. However, the concept of institution is one of the central concepts in social theory used in, for instance, Emile Durkheim's sociology. Today, there is a range of definitions of institutions in the diverse literature employing the term. The British anthropologist Radcliffe-Brown (1958: 174) defined an institution as '[a]n established or socially recognized system of norms or patterns of conduct referring to some aspect of social life'. Douglas (1986), another noted anthropologist, makes the argument that institutions are what structure human thinking into categories, defining what entities or events are similar or different. Douglas (1986) thus claims somewhat provocatively that institutions are capable of thinking. Adorno (2000: 105) emphasizes that institutions derive from and simultaneously guide action; institutions are then 'congealed action' – 'something which has become autonomously detached from direct social action'. Institutions are therefore, in Canguilhem's (1989: 380) formulation, a 'codification of a value, the embodiment of value as a set of rules'. Institutions are in short the bearers of social values and beliefs, and are therefore central to the understanding of social action. Czarniawska (1997) defends such a view of institutions:

> We cannot understand human conduct if we ignore its intentions, and we cannot understand human intentions or ignore the settings in which they make sense (Schütz, 1973). Such settings may be institutions, sets of practices, or some other contexts created by humans – contexts that have a history, within which both particular deeds and whole histories of individual actors can be and have to be situated, in order to be intelligible.
>
> (Czarniawska, 1997: 12)

To examine institutions and how they are functioning and influencing social action is therefore to understand both the 'doer and the deed'. Institutions are both embodying past actions and guiding and directing future action; they are links between the past, the present, and the future. More recently, the concept of institutions has been used in the management studies and organization theory literature. In a frequently cited paper, Barley and Tolbert (1997: 96) define institutions accordingly: '[W]e define institutions as *shared rules and typifications that identify categories of social actors and their appropriate activities of relationships*' [emphasis in the original]. This is a somewhat complex definition that needs to be unpacked. Using Schütz's (1962) term *typification* underlines that institutions are the outcome from repetitive social action; shared rules are what may be called sub-sets of institutions, the smallest analytical category when examining institutions. Typification means that social actors inscribe semi-stable qualities into events, entities or other actors. Similarly, rules are stabilized, 'congealed action'. Institutions are then what render what is fluid and contingent fixed and manageable. An additional and more recent definition is presented by Lanzara and Patriotta (2007: 637): 'We conceive institutionalisation as the phenomenological process by which a social order, a pattern, or a practice, comes to be taken for granted and is reproduced in structures that are to some extent self-sustaining'. This shares many components with Barley and Tolbert's (1997) definition. Again, institutions are what provide, with Giddens' (1990) term, 'ontological certainty' to a social practice; social actors do not need to doubt whether or not they partake in legitimate social action because their action is already sanctioned by the institution. It is precisely this layer of ontological certainty that the ethnomethodological studies advocated by Harold Garfinkel (1967) seek to destabilize and render problematic. Nevertheless, the concept of institutions has been used to inform organizational analysis during the entire post-World War II period. For instance, Philip Selznick (1957: 28) introduced the term *institutional leader* to denote leaders that are 'primarily an expert in the promotion and protection of values' [original emphasis omitted]. In their seminal work on 'resource-dependence' in organizations, Pfeffer and Salancik (1978: 234) invoked the term *institutionalization* when speaking of 'the establishment of relatively permanent structures and policies which favour one subunit's influence' in an organization. Other applications of the term exists in technology studies where, for instance, MacKenzie (1996) speaks of a *technological trajectory*, the developmental path of a specific technological artefact, as an institution:

> A technological trajectory is an institution. Like any institution, it is sustained not through any internal logic or through intrinsic superiority to other institutions, but because of the interest that develops in its continuance and the belief that it will continue. Its continuance becomes embedded in actors' framework of calculation and routine behavior, and it continues because it is thus embedded.
>
> (MacKenzie, 1996: 58)

In addition to the concept of institution, a series of terms have been used to capture the semi-stable nature and essentially predictable nature of social action in organizations. Such terms include *routine, rule* and *script*. First, routine is a central term in the organization theory and management studies vocabulary (for an overview of the literature, see Becker, 2004).

In the evolutionary theory perspective presented by Nelson and Winter (1982: 14), routines 'play the role that genes play in biological evolutionary theory'. That is, routines are 'persistent features' of the organization; they store information and know-how, and they convey information from experienced co-workers to newcomers. Routines are, in brief, a central component of the organization's memory. More recently, Martha Feldman and colleagues have examined the concept of the routine in a number of publications (Feldman, 2000; Feldman and Rafaeli, 2002; Feldman and Pentland, 2003, 2005). Feldman (2000) argues that, rather than being fixed and petrified, a conventional view of the routine shaped by a scepticism towards bureaucratic organization forms, routines are flexible scripts that enable organizations to change and restructure. Feldman and Rafaeli (2002: 311) define routines as 'recurring patterns of behaviour of multiple organizational members involved in informing organizational tasks' [original emphasis omitted]. In a more recent paper, Feldman and Pentland speak of routines in the following terms:

> We define organizational routines as repetitive, recognizable patterns of interdependent actions, carried out by multiple actors. We claim that organizational routines combine an ostensive aspect, the ideal or schematic form of a routine and a performative aspect, specific actions by specific people in specific places and times. Any particular routine within an organization can be analysed in terms of these parts and the interactions between them.
>
> (Feldman and Pentland, 2005: 96)

All routines are therefore dual: one part is abstract, ideal and ostensive and represents the structural component of the routine; one part is the actual performance of the routine. This separation between the idealized and formal, and the actual and practised, corresponds to what Ferdinand de Saussure (1959) referred to as *la langue* and *la parole*, the formal structure of language and its actual use in real life setting. The reliance on semiotic theory is recognized by Pentland and Rueter (1994) explicitly stating that routines are the 'grammars of action'. Speaking in such semiotic terms, Feldman and Pentland (2005) remark that all routines require both components and that any analysis of organizational routines needs to take into account both components. Seen in this view, routines are, in analogy with language, 'both stable and adaptable at the same time' (Feldman and Rafaeli, 2002: 325). Seen as a central mechanism for many parts of social life – everyday life expressions use many references to personal or collective routines like 'my morning routine' or 'routine work' – routines have been critically discussed as what are underlying monotonous and repetitive everyday work-life experiences, thereby being cast as what is problematic *per se*. While routines can demean (Sennett, 1998: 43), they can also protect social actors in various working-life situations (Leidner, 1993); for instance, referring to policies and routines is an effective argument against excessive expectations and claims from customers.

Second, rule is a concept closely related to routines, but has more formal and prohibitory connotations. Tsoukas (2005: 75), speaking from a knowledge-management perspective, emphasizes the universal and the particular, similar to Feldman and Pentland (2003, 2005) in their analysis of routines, that rules are a link between 'general categories' and 'particular instances'. However, as any manager will tell you, this link is far from perfect; there is a precarious connection between the formal prescriptions and the actual performances, and what determines the smooth integration of the two levels of the rule is experience, skills and detailed know-how of the rule. That is, rules are formal scripts guiding the social actor in his or her activities. Tsoukas (2005) emphasizes that rules are in most cases accompanied by narratives, stories informing the newcomer on how to deal with the practical matters prescribed in the rule. 'Rules cannot have the role that narratives have: rules are impersonal, generic, and atemporal formulae bearing only an apparent relation to what I am exactly experiencing on "the ground"', Tsoukas (2005: 82) argues. Thoughtful action is then the ability to navigate between formal rules and effective actions, and the seasoned co-worker may tell stories on how to accomplish such skilled performances. Narrating social action is one of the principal means for learning in organizations (Orr, 1996).

Third and finally, some researchers have used the term *script* to denote how knowledge is instituted in organizations. For Barley and Tolbert (1997: 98), '[s]cripts are observable, recurrent activities and patterns of interactions characteristic of a particular setting' [original emphasis omitted]. Scripts are thus capable of encoding 'the social logic' of what Erwin Goffman called an 'interaction order'. Powell (1985: 148), making a reference to what he calls 'socialization theory', argues that the person's 'social self' and the behaviours of others influence how he or she will act out the script, but the script nevertheless determines most of the appropriate behaviour for a certain social position. Scripts are thus somewhat flexible without being negotiable *per se*. Timmermans and Berg (1997) study how a particular form of script, a protocol, was used to organize and structure the work in the emergency ward in a Dutch hospital. In their view, a 'technoscientific script', that is, a script derived from a setting dominated by the gradual convergence of scientific skills and thinking and accompanying technology, is capable of specifying 'actions, settings, and actors who are defined with specific tastes, motives, aspirations, political prejudices, and a value system' (Timmermans and Berg, 1997: 275). The role of the script is thus to coordinate and align a series of actions, practices, tools and technologies in medical practice. In the emergency ward, fast and goal-oriented action is pivotal for the quality of the health care provided and the script serves to normalize adequate action into scripted activities.

Timmermans and Berg (1997) claim that the script enables what they call *local universality*, meaning that generally applicable principles are constituted on the basis of local and particular actions. Universality is then a 'non-transcendental term' no longer implying 'a rupture with the "local", but transforming and emerging in and through it' (Timmermans and Berg, 1997: 275).

Managerial action in institutionalized settings

In 1934, Adolf Berle and Gardiner Means published their landmark study *The Modern Corporation and Private Property*, wherein they addressed the increased reliance on professional managers in the United States' economy. While the previous dominant corporate governance regime relied on what Fligstein (1990) calls 'the direct control of entrepreneurs', the owners of the corporations, in the interwar period a class of professional managers – a class of white-collar workers meticulously examined by sociologists such as Charles Wright Mills (1951, 1956) and William H. Whyte (1956) in the 1950s – were made responsible for the control and maintenance of corporations. The new managers were, contrary to the entrepreneurial class, paid on the basis of a fixed salary and were therefore primarily interested in the survival of the corporation rather than to maximize profits and exploit opportunities. Fligstein (1990: 3) emphasizes that managers are not, contrary to what is assumed in neo-classical economic rational-choice theory, 'maximizers' or even 'satisficers' but rather 'behave to preserve what is'. Fligstein (1990), critical of abstract theories regarding the behaviour of actors such as managers, argues that the individual manager's view of the world and what actions to take is largely determined by his or her position in the organization, which in turn 'form the interests and identities of actors' (Fligstein, 1990: 11). Seen in this view, that is, in an institutional perspective, the actions of accountable managers are largely determined by external conditions and established norms and values.

Meyer and Rowan (1977) argue that organizations are 'driven' to incorporate practices and procedures institutionalized in society. That is, organizations maintain 'ceremonial conformity' and reflect institutional rules. This conformism or 'isomorphism', the incorporation of elements that are 'legitimated externally', are not irrational but help the organization to reduce turbulence, maintain stability and handle uncertainty (Meyer and Rowan, 1977: 349–9). DiMaggio and Powell (1983: 150) identify three forms of isomorphism: (1) a *coercive* isomorphism that 'stems from political influence and the problem of legitimacy'; (2) enact a *mimetic* isomorphism resulting from 'standard responses to uncertainty'; and (3) a *normative* isomorphism, associated with professionalization. Ultimately, the conformism to institutions promotes long-term success and survival for the organization. One of the consequences of the adherence to instituted norms and practices is that organizations tend to, Meyer and

Rowan (1977) argue, 'disappear as distinct and bounded units'. They also remark:

> Quite beyond the environmental interrelations suggested in open-systems theories, institutional theories in their extreme forms define organizations as dramatic enactments of the rationalized myths pervading modern societies, rather than as units involved in exchange – no matter how complex – with their environments.
>
> (Meyer and Rowan, 1977: 346)

Meyer and Rowan (1977) emphasize that instituted norms and practices, brought into the organization to gain legitimacy, do not need to be rational, effective or capable of dealing with organizational concerns. Instead, there may be significant conflicts between 'categorical rules' and 'the logic of efficiency', thereby forcing organizations to either: (1) decouple formal structures and actual activities; or (2) institute a 'logic of confidence and good faith'. This means that organizations are 'hypocritical' (Brunsson, 1985) and establish what Argyris and Schön (1978) speak of as the separation between *espoused theories* and *theories-in-use*; organizations, aiming at reconciling opposing or contradictory objectives, are therefore doing one thing and saying another (see e.g. Dalton, 1959; Jackall, 1988). 'Organizations often face the dilemma that activities celebrating institutionalized rules, although they count as virtuous ceremonial expenditures, are pure costs from the point of view of efficiency', Meyer and Rowan (1977: 355) note. What Meyer and Rowan (1977) call the 'logic of confidence and good faith' means that organization members are confident of the rationality in the actual practices and routines. Although it may be complicated to justify in financial terms the hiring of a consultant or making an investment in training on the basis of economic returns, there is a common belief in the value of adhering to such instituted practices in organizations.

In Berle and Means's (1934) pioneering discussion on the effects of the emerging 'managerial' regime of corporate governance, there is a major concern regarding the effectiveness of the new managerialist regime. Institutional theory largely confirms such worries in emphasizing the 'myth and ceremony' of co-optation with the environment, the various forms of isomorphisms in organizational fields such as specific industries. At the same time, institutional theory also underlines that efficiencies are achieved through a number of mechanisms and procedures reconciling such conflicts of interests (Dalton, 1959). Speaking of institutions in a knowledge-management setting therefore means not to assume beforehand that organizational routines and practices for developing, sharing and preserving knowledge are established solely on the basis of an efficiency objective. Instead, knowledge-management practices could be adopted on the basis of their degree of legitimacy in the industry rather than their practical value.

Instituting knowledge

Lanzara and Patriotta (2007) study how knowledge is instituted in a newly built Fiat factory in southern Italy. They suggest that the institutionalization of knowledge is what makes knowledge stable and unified, makes knowledge less susceptible for local interpretations and controversies and thereby lowers the demand for accompanying social control. When knowledge is highly instituted, it is accepted without disputes and has a high degree of authority among relevant actors. Institutionalization is what helps enact social practices and renders them uncomplicated, stable and predictable. Patriotta (2003) here speaks of the production of 'epistemological closure' wherein 'emergent stocks of knowledge' are 'sealed' (see also Bechky, 2003a). However, such epistemological closures are susceptible to revisions, especially in periods of turmoil when, for instance, new technologies are introduced, conflicts between groups emerge, or during periods of radical change in the organization or in its environment.

The process of institutionalization is by no means an uncomplicated or linear process. Lanzara and Patriotta (2007: 638) propose the concept of *template* to denote the 'generative principle' or 'code' that reproduces a behavioural pattern across a variety of media, artefacts and organizational devices. A template can then be defined as a practical example, often based on 'shared cognitive analogy' that becomes the commonly accepted way of doing things. The template is, Lanzara and Patriotta (2007) say, 'a master model or pattern by which other similar things can be made'. Templates are both generative in terms of enabling further action, and normative in terms of providing guidelines and standards for how operations are to be conducted. Patriotta explicated this idea:

> As a result of institutionalisation, knowledge is inscribed into a system of norms, practices and conventions, and incorporated into stable structures. Knowledge becomes canonical, factual, definite and certain. Institutionalization implies a process of epistemological closure similar to the closure of black boxes.
>
> (Patriotta, 2003: 181)

Social action is structured in accordance with templates and templates are the generative code for further action, guiding and directing practice. The perspective on institutionalization presented by Lanzara and Patriotta (2007) and Patriotta (2003) thus assumes a recursivity between structure (template) and action that shares many qualities with Giddens' (1984) *structuration theory*; social structure and social action are never isolated entities or events, but have a dynamic and intrinsic relationship – they are mutually constitutive. Lanzara and Patriotta's (2007) and Patriotta's (2003) objective is, however, unlike Giddens, not to present a formal theory of society but to examine how knowledge is instituted and reproduced in an actual organizational setting. In their view, knowledge is what is gradually instituted but what is at the same time always susceptible to disputes and controversies. That is, 'knowledge systems' (Dougherty, 2007) are always open for debate and controversy. For instance, in Fujimura's (1996) study of so-called oncogene research, one of the most widely acclaimed research programmes in the 1980s and 1990s in oncology research, the success of the oncogene research depended on a variety of factors. For instance, the oncogene approach enabled the articulation of what Fujimura calls 'doable problems', that is, the approach offered opportunities for grappling with actual medical problems without inventing new technology, abandoning predominant beliefs and radically reconsidering the domain of expertise. In addition, the enthusiasm for the new approach – Fujimura (1996: 3) speaks about jumping on 'the oncogene bandwagon' – derived from the general belief that the oncogene framework was not threatening any other pre-existing domain of research and expertise:

> [T]he proto-oncogene theory did not challenge the theories to which the researchers had made previous commitments. Indeed, the new research provided them with ways of triangulating evidence using new methods and a new unit of analysis to support earlier ideas. These view of oncogene research were 'realized' through the efforts of these researchers and, in turn, this realization further extended the reach of oncogene research and the complexity of the theory.
>
> (Fujimura, 1996: 151)

The institutionalization of the oncogene theory approach was supported by the absence of controversies, and relatively little inertia and resistance in the community of medical researchers. Compared with Lanzara and Patriotta's study of Fiat, both Fujimura and the study of the Italian automotive manufacturer underline the importance of shared social values and norms embedding know-how and skills.

In summary, one must not assume that knowledge exists *per se* in organizations, or that knowledge is immediately recognized and well received in social groups or communities. Instead, knowledge is always already shaped by predominant beliefs and assumptions. The institutional perspective on knowledge management teaches us carefully to take into account the ideologies and objectives of participating actors in a specific field. Without their compliance and recognition of new forms of knowledge, most new insights fall flat. The institutionalization of new knowledge therefore shares many characteristics with new technology adaptation. As, for instance, Barley (1986, 1990) has shown, new technology redefines social relations and therefore is often controversial. In Barley's (1990) account:

> Technologies are depicted as implanting or removing skills much as a surgeon would insert a pacemaker or remove a gall bladder.

Rarely, however, is the process so tidy. Events subsequent to the introduction of a technology may show that reputedly obsolete skills retain their importance, that new skills surface to replace those that were made redundant, or that matters of skill remain unresolved. In any case, groups will surely jockey for the right to define their roles to their own advantage.

(Barley, 1990: 67)

The same argument is valid for new knowledge; new domains of knowledge are often regarded as threatening or, alternatively, as being an opportunity for various groups of social actors. Therefore, the use of new knowledge is a complex social process that needs to be examined as such.

Thinking of knowledge as what is instituted in organizations means to recognize that knowledge is always already distributed over communities, teams, practices, tools, technical systems and so forth. Knowledge is then always in-the-making at the same time as it is grounded in the semi-stable temporospatial structures that are referred to as institutions. The institutionalization of knowledge means that it is part of a broader set of practices and routines that will promote the long-term survival of the know-how in question.

The institutionalization of knowledge in the construction industry: the four case studies revisited

Following Lanzara and Patriotta (2007), using the concept of a template to denote the 'generative principle' or 'code' that reproduces a behavioural pattern across a variety of media, artefacts and organizational devices, the four cases of knowledge use in construction industry will be discussed in this section. Institutionalization is here the process wherein knowledge is rendered not only legitimate and practically useful but also helps define what is a proper practice within a specific field. As suggested by Strauss, Schatzman, Bucher *et al.* (1964), medical doctors in the domain of psychiatry did not only adhere to specific treatment forms for the patients but the instituted regimes of knowledge also provided clear normative and moral grounds for the joint work in the hospital or community of medical doctors. Instituting knowledge is then not only to establish knowledge as a resource but also a way to make co-workers comply with certain standards and objectives. It imposes a veritable *modus operandum*, a 'one best way' of doing things.

In Chapter 2, site managers' work was examined as playing a central yet vulnerable role in the construction industry. Numerous practitioners and researchers have emphasized that site managers are carrying a significant workload and take on the burden of being responsible for a wide variety of domains of expertise, ranging from hands-on day-to-day communication with the co-workers to legal matters and environmental issues. The literature also suggests that site managers are operating in a rather homogeneous community, dominated by white, working-class men, characterized by a culture wherein each and everyone is responsible on his own for his work and his accomplishments. Site managers are expected to represent the 'lone wolf' personality, self-sufficiently making decisions and acting as the leading man at his construction site. It is, in other words, a highly gendered culture.

The study shows that site managers appreciated the coaching support, and thought of executive coaching as an adequate means to institute new means of learning and developing new skills in a real working-life setting. As opposed to more conventional management training courses, the coaching programme was centred on the site managers' personal experiences and concerns rather than ideal typical illustrations collected from textbooks designed to enable class room training. In addition, the coaching programme encouraged the site managers to actively use new insights and experiences in the work-life setting. There was a rather short distance (both spatially and temporally) from the discussions and conversations with the coach every second or third week to the actual work-life setting in comparison with conventional management training. Speaking in Lanzara and Patriotta's (2007) terms, the coaching process is a template for the management of knowledge in the construction industry, a generative principle that both helps to share knowledge and to produce new knowledge in a systematic manner; the coaching practice, constituted by the very event of coaching, the time period the coach and the coachee, in our case a site manager, spend together is what helps institutionalize knowledge. First, the very act of the coaching as a formally sanctioned event, wherein a co-worker in the construction industry articulates and elaborates on his or her concerns and forthcoming challenges, serves to institutionalize knowledge-management procedures. Here, a site manager is stepping out of the conventional site manager role as being 'a doer' or a 'man/woman of action', and instead actively reflects on the pros and cons of previous and predominant action. The coaching process then provides an arena where site managers can think of themselves in new terms and construct new images of themselves and their professional roles. Seen in this perspective, the coaching process aims at unearthing a whole set of assumptions and beliefs that are taken for granted regarding the site managers' role and the nature of the construction industry as such. Second, the coaching process institutes new knowledge in terms of imposing demands (or at least suggesting that it may be helpful to accomplish more effective site-management work) for not only reflecting on practices, but also actively modifying and altering an individual's behaviour in the site management role. Site managers are, in other words, not only expected to examine their own and collective actions in the workplace critically, but also to let such analyses shape and inform further action actively. The coaching process is thus instituting what has been called the 'plan–check–do–act' learning cycle (at times also called the Deming cycle) that has been suggested in the organization learning literature and the operations management literature. Taken together, executive coaching institutes knowledge in terms of imposing the normative position that site managers should actively examine their own behaviour and alter it to accomplish the best possible outcome. While such beliefs may be widely subscribed to in general, the coaching practice is the very social practice in which such beliefs are actively transformed into social practice. Site managers and coaches are in fact meeting and discussing site managers' work and site managers are actively encouraged to modify their behaviour in work-life situations. The coaching process is instituting new knowledge by continuously alternating between reflection and action.

In the case of architects' work, the process of institutionalization is more complex and variegated. Architect work operates in the boundary between the material and the symbolic, is embedded in talk and communication in its day-to-day functioning, draws on external influences and ideas developed outside of the focal firm, and gets its impetus from peer recognition and peer reviews. The materialized semiosis of architect work is an epistemological and analytical facet of architect work, suggesting that knowledge is instituted in communities of architects in the form of sketches, models, images and blueprints of future buildings. Similar to scientific communities, elaborating with 'epistemic objects' that are too material to be wholly abstract and too symbolic to be a specimen of an undisputed *de facto* nature, architects are bridging aesthetic, practical, political and material aspects of the forthcoming building in their models and sketches. Similar to the role played by theories, hypotheses and theorems in scientific communities, guiding and coordinating research interests and research activities, architects use models and sketches as boundary-objects, bringing heterogeneous groups into meaningful conversations. In the vocabulary of institutionalization of knowledge, models and sketches are not mere heuristics 'additional' to the architect work but are instead a principal means for instituting knowledge. For instance, it is commonplace that architecture magazines and periodicals publish models and sketches after the building is finished to show how the thought processes evolved during the project. Second, talk and communication is another principal means of institutionalizing knowledge. Architects continually exchange ideas, remarks on sketches, comments on one another's suggestions and pass judgment on various architectural works. The very community of architects is maintained through its active conversation. Externally, architects must learn to master a vocabulary that promotes their interests, yet are capable of sounding credible and reasonable for non-architects. Again, similar to scientists, expected to deliver practical knowledge to the broader social community, architects cannot operate in self-enclosed environments but are always required to take into account a wide range of issues and concerns in their work. Talk and communication is therefore of central importance to the discipline and is thus of interest for active management and promotion from a managerial perspective. Third, architects have an active exchange with the broader community of architects and individuals or groups of individuals interested in architecture. Most new ideas but also the recognition and status within the field of architecture derive from the outside. One way to actively promote and institutionalize new knowledge in architect work is thus to encourage the reading of architecture periodicals and architecture press, to participate in industry events and to pay attention to what is happening in the field in general. In scientific communities, knowledge is acquired through following a few peer-reviewed journals central to the discipline and through participating in scientific conferences and congresses. State-of-the-art researchers in fact invest a significant amount of time to stay updated in their field of expertise. Especially in the natural sciences and the so-called life sciences, the research frontier is advancing fast. The reputation of individual architect work and architects is a resource of great value when attracting new and returning customers. However, as all marketing analysts and marketers know, it is also a notoriously complex social process to manage taste and reputation. Architects' firms and architects may, however, choose certain strategies on how to develop a strong brand in the industry; for instance, through focusing a specific type of building or targeting prestigious competitions they know will gain substantial attention and recognition in the field.

In general, study of the Blue Architect Firm shows that the institutionalization of knowledge operates on many levels and includes many social practices. Knowledge, it may be suggested, is primarily institutionalized through professional identification wherein individual co-workers and newcomers actively subscribe to various kinds of professional ethos. What counts as legitimate knowledge is then not in the first place embedded in its actual use-value, but its ability to be aligned and brought into harmony with predominant beliefs and norms regarding the nature of architect work. In other words, to manage knowledge in architects' offices is to manipulate professional cultures and professional identities; instituting knowledge demands a change of perspective from what is widely agreed upon to new images of architect work. As in all professional settings, knowledge is not what is additional to professional identities but is instead what is intimately bound up with such beliefs and ideologies. There is then no detached or isolated plane of knowledge being added on top of the professional plane; knowledge and professional identities are entangled in various ways.

The study of Brown Architects, the third study accounted for in Chapter 4, suggests that architects' know-how and competence do not only reside in the cognitive capacities of the practising architects. Instead, the architects' competencies and skills are instead bridging material and concrete resources and the abstract and intangible values and norms that are based on the aesthetic knowledge of the community of architects. The ability to examine a programme, the written specifications provided by the client, and to translate these specifications into an architectural solution that is taking into account practical, material, financial and aesthetic considerations, is a competence that demands a certain capacity of seeing the opportunities of the particular programme and the specific site. Through training and practical experience, architects develop a professional vision, an architect's gaze that embodies and include all of the architect's know-how and experience. Seen in this view, the knowledge of the architect is no longer what is solely located in the capacity for human cognition, but is instead an embodied and distributed resource; embodied in terms of being what is anchored in the professional vision, the perceptual skills of the practising architect; distributed in terms of being what Lacan calls 'the gaze' [French, *le regard*], the look that is always already interpenetrated by the look of the other. While the look in the Lacanian view is a concept presupposing that ideologies are always present in the ways of seeing, in the concept of the architect's gaze, there are no such inherently negative connotations. Instead, the architect's gaze is a manifestation of collective or communal knowledge, knowledge that is acquired through months and years of inspecting, evaluating, discussing visual artefacts such as photos, computer-aided design (CAD) images and models with colleagues and peers. The architect's gaze comprises the totality of ways of seeing enacted within the community of architects. Speaking of institutionalization of knowledge, managers need to support the joint use of visual artefacts, and to nourish a continuous debate and discussion about architecture in the firm. For instance, in Brown Architects, seminars and excursions were organized and the co-workers were given the opportunity to discuss built architecture on site. For architects, the experience of actual built spaces is of central importance for their identity and their professional skills. Without the ability to 'look, smell, touch and feel' actual buildings, their collective competence is arguably hollowed out over time. Instituting knowledge work in architect firms therefore includes paying attention to the professional vision of the community of architects. In very practical terms, this may be translated into arranging seminars and workshops, investing time and resources in excursions and in general providing a variety of visual tools and artefacts (i.e. relevant journals and magazines, software and other media) that are useful in the early phases of the architect's work process, the 'fuzzy front end' where the programmes, ideas and concepts are being shaped by creative thinking. Moreover, innovative and qualified architecture demands that there are adequate time frames to allow for some creative thinking. In addition to managerial practice that supports the professional vision of the community of architects, more regular procedures for sharing insights and experience 'upstream and downstream' of the construction process are pivotal for long-term competitiveness. The entire construction industry may benefit from more rigorous and solid knowledge-sharing and project termination procedures. At least, most industry representatives point to this as a perennial issue in the industry.

A fourth study of the specialized construction company expert in rock construction work and, more specifically, sprayed concrete, rock injection and rock reinforcement, points at two aspects of construction work. First, social capital is an important organizational resource when it comes to knowledge-sharing in communities that are not overtly interested in formalizing their collective know-how. Such communities are primarily oral, that is, they rely on their ability to share and circulate knowledge through oral communication, either face-to-face or over the telephone, and conceive of written communication as being too time-consuming to be justified. In oral communities, the increased use of new media (i.e. computers, databases, the internet) in society is partially accommodated, but it is nevertheless the ability to express insights and know-how verbally that makes the largest difference in day-to-day work settings. This emphasis on oral communication is by no means the result of sloppy management procedures but instead derives from the idiosyncrasies of the industry. For instance, in ConCo, formal reporting and evaluation of terminated projects were generally regarded to be too time-consuming and, in addition, some projects were not expected to be returning within the next ten years anyway. Therefore, the ConCo site managers were not willing to spend additional time writing extensive documentation reporting how they proceeded and what they learned from the case. Instead, the ConCo site managers actively promoted a professional ideology, wherein the site manager's role is not in the first place to predict and anticipate upcoming challenges but rather to be capable of solving problems as they occur. Predicting site construction work includes too many variables and events, the site managers argued, but being able to find solutions to problems is quite a different matter. Speaking in terms of institutionalization of knowledge, social capital is the shared and relational capacity to address and solve such site management specific problems collectively. Social capital is, as pointed out by a number of sociologists and organization theorists, intangible and is not controlled by individuals but rather emerges in the joint collaborations between individuals. Yet, social capital is productive in terms of contributing to accomplishing certain social objectives, in this case, matters relating to construction sites. Institutionalizing social capital then encourages individuals to speak openly to one another on topics pertaining to their work assignments and to promote an organizational and professional culture, wherein individuals both contribute and draw on existing stocks of knowledge.

Second, the study of ConCo shows how even work fairly well removed from the production of ornamental or aesthetic products may be fruitfully examined as being a social practice, drawing on what has been called 'aesthetic knowledge'. The work of the ConCo construction workers took place in tunnels and underground systems constituting the infrastructure of modern cities and societies. Seen in this way, their work was not concerned with conventional aesthetic work. Yet, their individual competencies and skills were addressed by the ConCo site managers in a vocabulary that only vaguely related to the logical–rational narrative that engineers and construction workers are prone to use in many cases. Instead, the site managers talked about the construction workers' skills in terms of their ability to integrate a series of human faculties and skills. For instance, the best spray-concrete robot operators were capable of listening to machinery to tell how it was operating, of producing aesthetically appealing works and of taking into account a wide range of aspects in their work. These elite robot operators were not only pursuing their work as a series of detached activities, largely independent from one another, but could bring these various operations into a functional unity. Their ability to bridge machinery, the concrete and other material resources, and the actual operation was a highly praised skill in the site management community. In terms of institutionalization, aesthetic knowledge is, by definition, very complicated to share and transfer, and is to a higher degree based on tacit rather than propositional knowledge. Therefore, institutionalizing aesthetic knowledge is a complex social undertaking, strongly emphasizing learning by doing and 'learning by watching' as important processes in vocational training. However, as the CEO of ConCo emphasized, it is always tempting for the site manager to use the best operator to do the same job over and over, thus to some extent blocking the opportunities for newcomers to acquire new skills and competencies. However, the site managers were aware of such challenges and actively promoted novices to train for new skills. Institutionalizing aesthetic knowledge is thus based on the idea of practical training, to engage in repetitive work in a specific domain.

The four case studies suggest that there is a considerable difference between the various professions and activities located in the construction industry; social practice emerges in many forms and in many settings and therefore the value of an abstract theoretical framework lies in its ability to make sense out of and systematize what is idiosyncratic and local. To look at the consequences from this insight, there will be no attempt at formulating a universally valid 'theory' about the institutionalization of knowledge in the construction industry. Instead, it is recognized that the various uses of knowledge in the industry are based on what Timmermans and Berg (1997) call 'local universality'; they are locally enacted and applied but with the intention to operate universally and to be representative of a generalized practice. The institutionalization of knowledge in a local universality setting is emergent, contingent and context-bound. Seen in this way, there is no 'one size fits all' theory available here. The construction industry is not a unified and integrated whole, but is more of a patchwork of professions, firms, technologies, materials and so forth. Managing knowledge in such a patchwork is therefore, in the first place, to pay attention to how local cultures, contingent professional ideologies, and established social practices are developing, shaping and circulating bodies of knowledge. Following Sole and Edmondson (2002), knowledge is always 'situated', embedded in cultural, social, material and cultural settings that strongly determine what counts as knowledge and what does not. Knowledge is in this view not primarily what exists *per se*, that is, as a self-enclosed entity, a set of propositions ready to apply to cases, but is instead what is 'in-the-making' (Patriotta, 2003), that is, a social accomplishment that is the outcome from the collaboration between heterogeneous groups. 'Knowledge is nothing without a collective to found it', Michel Serres (1991: 104) claims. This is a declarative statement for a knowledge-in-the-making perspective, an insistence on thinking of knowledge not as an entity but as a process, a flow, a social practice including a variety of resources and materials.

Implications for managerial practice in the construction industry

Even though the analysis of the various knowledge-management practices in this book concludes that one must not make what Ryle (1949) calls a 'category mistake', to believe that knowledge is what can be seen, touched and smelled as such, but what is rather being distributed in the polymorphous forms of social action that constitute everyday work-life in the industry, there is, arguably, an obligation to articulate some practical implications from the three case studies. Organization theory and management studies is, in comparison with the liberal arts and the natural sciences, a plebeian science; it is a science (or rather, a set of inter-related theories) that deals with the nuts and bolts of everyday life in all its mundane activities. It is a science addressing bookkeeping, management control, distribution activities and so forth, not being removed from the everyday work-life of millions of practising managers and co-workers in organizations. However, no matter what credentials and affiliations an author may invoke to gain authority in a field of expertise, presenting advice for practising managers is always a risky business. Still, some thought may deserve to be expressed here.

First, the four cases point at the need for examining the knowledge used in everyday practice in the construction industry in its social context, a social context often filled with conflicting goals, timelines imposing a demand for giving priority to certain activities, a certain degree of ambiguity, and a more or less explicit conflict of interests between individuals or groups of individuals. The ability of construction industry co-workers to 'make things work' is the capacity of mobilizing know-how in the middle of this hodgepodge of practices, materials and beliefs. From a managerial perspective, knowledge must not be treated as some kind of tool that can be deduced and applied to cases. Some propositional knowledge may serve this function, especially knowledge that deals with juridical matters or industry standards. For instance, a familiarity with building standards and norms is of central importance for the site managers in ConCo unless they want to be challenged by the client representatives when the work is being examined. In most cases, however, knowledge is more vague, fuzzy and contingent, and helps the site managers run their projects effectively on the basis of a set of skills and experiences that are at best partially articulated. To manage this category of knowledge, managers then need to be able to understand the nature of knowledge as such but also the very 'context of application', that is, to understand how communities of various professional and occupational groups jointly collaborate to accomplish certain objectives. For instance, in ConCo, what has been referred to as 'social capital', the ability and willingness to share insights and experiences with one's colleagues, played a central role when sharing and distributing knowledge. The expertise of ConCo lies, arguably, not in the totality of the individual site managers' skill and know-how but in their ability to share this stock of expertise through verbal communication. Managing knowledge in the construction industry is then to see how knowledge is of necessity anchored in social relations.

Second, to manage knowledge in the construction industry may necessitate taking a strong position in favour of certain practices, values or norms. For instance, in-site management work, the very conventional view of how site managers should act, behave and think of themselves and their relationships to colleagues and superiors, may be challenged and called into question for the benefit of new thinking. Site managers endure a rather restrained work-life situation and additional administrative workloads are further burdening them. In such a situation, the established and taken-for-granted views of the position of the site manager might be de-familiarized and contested. The study of the use of coaching practices to support site managers in their operations suggests that the coach helped the site managers to articulate a new perspective on the individual's competencies and his work; the coaching process opened up the 'black box' of site management and helped call attention to internal contradictions or tension in the very site manager role. When managing knowledge in the construction industry, there may be a need to institute and establish mechanisms that de-familiarize and contest all-too-naturalized views of how the industry should operate, and how different professional and occupational groups should preferably relate to one another. Managing knowledge is not a trivial matter but includes a variety of emotional, cognitive and aesthetic human skills. When knowledge work is solely embedded in routines and fairly rigid scripts, individuals may lose their interests and demonstrate stereotypical behaviour. Managing knowledge implies an ability to uproot this 'everydayness' of social action and call attention to new ways of conducting the work. For instance, enacting a social practice, such as an executive coaching programme, is one example of an organized activity that helps an occupational group (in our case, site managers) conceive of their work in new terms and from new perspectives.

The management of knowledge is therefore, again, to encourage a critical self-reflection of work practices and routines among the co-workers.

Third, the construction industry is, especially in the production phases, dominated by oral rather than written or mediated communication, that is, knowledge is primarily shared through storytelling and narrative accounts of experiences in the day-to-day work and not in lengthy written reports. Managing knowledge in such 'oral cultures' requires the provision of shared arenas where actors can meet and articulate their learning and experiences. In many cases, the distribution of such verbally transmitted know-how is expected to occur all by itself, uninterrupted by managerial initiatives. However, many studies of communication in organizations show that oral communication is susceptible to time and space limitations; normally individuals in organizations primarily speak to people in their own office corridor, their own departments, or in the project wherein he or she works for the time being. A managerial implication is that relying on know-how to be distributed through such informal channels may fail to exploit the full potential of the accumulated knowledge in a company. Some examples of active engagement in supporting the sharing of know-how include the knowledge facilitation model developed by Jonas Roth (2003), employed to share know-how across project boundaries in new drug development projects. In the so-called 'knowledge facilitation model', project teams meet in a two-hour seminar moderated by a so-called knowledge facilitator and jointly articulate what they have learned during the process. Research suggests that such knowledge facilitation seminars save considerable amounts of money in the clinical trails in the new drug development process. In the construction industry, site managers are often expected to act as 'enterprising selves', to be able to handle all their activities on their own and without top management involvement. While this conventional view of the site manager may have some merits, there is also a risk in leaving the site manager too detached from the line organization and his or her site manager colleagues. Instead, construction companies may organize seminars, meetings, lunches, etc., wherein recently acquired know-how can be jointly articulated and shared. The traditional '*laissez-faire* model' for knowledge-sharing demands little time from the construction industry co-workers but it is also a fickle model whose ability to exploit underlying knowledge bases may be called into question.

Fourth, a new vocabulary leaving the metaphor 'knowledge-as-stock' behind may be useful. Being a knowledge-intensive firm is to develop and use a conceptual and operative vocabulary capable of denoting the underlying skills, experiences and know-how in use. While knowledge may be regarded as a relatively fixed and self-enclosed category, it may have some merits and uses in some analytical endeavours (e.g. in political policy and in economic theory) but, in the day-to-day management of knowledge in construction companies, it is more helpful to conceive of knowledge as the outcome from the collaboration of co-workers under determinate conditions. Seen in this view, knowledge is not a 'tool' but a social accomplishment, embedded in heterogeneous assemblages constituted by practices, routines, ideologies, materials, blueprints and sketches, and other resources mobilized and brought into action in construction projects. Knowledge is not what is separated from the material resources but is rather what emerges in the active engagement with the construction of the building; know-how is acquired through practices but know-how and expertise also structure and organize practices. In the analysis of day-to-day work in construction companies, it is therefore not analytically meaningful to separate knowledge and practice into causal relations, but instead the two have a reciprocal and a recursive relationship wherein the one is always entangled with the other. Expressed differently, rather than assuming that knowledge is always already 'in place' and employing a vocabulary reinforcing such an image of knowledge, a complementary vocabulary may be helpful to conceive of new ways of apprehending knowledge in construction work. That is to say that the technical and engineering-based terminology of much construction work practice and research may be complemented by – what we may call for the lack of a more appropriate term – a more sociological or socio-cultural vocabulary emphasizing the social embeddedness of knowledge.

Besides these general recommendations, there may be a wide range of implications from the studies. However, such implications are likely to be dependent on local conditions and idiosyncratic professional and occupational ideologies structuring the day-to-day work in various parts of the construction industry. Avoiding the trap of playing the expert outlining detailed scripts for 'how to' accomplish various objectives, there will be no speculations about what these implications actually may be.

Summary and conclusion

This book has aimed at discussing the management of knowledge in the construction industry. Rather than assuming that knowledge is a fixed entity, a self-enclosed and largely uncomplicated social resource ready to apply to cases, knowledge is conceived of as what emerges in social action and in everyday practice. While such declarations may appear uncomplicated or relatively uncontested a priori, in the study of the day-to-day management of organizations, it demands an analytical approach that takes into account a broader set of factors than just the 'knowledge-in-use' *per se* (which is here an abstract term that may be permitted for the sake of making an argument). That is, knowledge is not the starting point for the analysis but is rather the *outcome* from the everyday practice; at the same time, there must be some kind of structured experience and entrenched know-how that guides social action – knowledge is therefore to some extent both the cause and the effect of social practice. In such a sociological perspective on knowledge, there is a recursive relationship between the actor and the structure, between knowledge and social practice; they are mutually constructive and are implied in one another. No knowledge without social practice, no social practice without knowledge. Speaking in more empirically oriented terms, the construction industry is a heterogeneous assemblage constituted by a variety of professions, occupations, materials, practices and so forth. To manage knowledge in such a diverse setting demands a variety of approaches and practices. For instance, architects, site managers and rock construction workers are by no means mobilizing and employing the same body of know-how, although their idiosyncratic expertise and know-how is intersecting at the actual construction site and in the everyday work-life to transform materials into buildings and civil engineering projects. One of the key lessons from the four case studies is that there is no such thing as a 'one size fits all theory' when it comes to knowledge management in the construction industry. Instead, equally analytical pursuits and practical day-to-day managerial work benefit from recognizing and apprehending the locally enacted and culturally embedded social practices which constitute knowledge-work in construction companies, architecture bureaus and at construction sites. Expressed differently, to manage knowledge is to manage social relations and social relations are in themselves embedded in the abstract analytical categories that social researchers refer to as institutions, professional and occupational ideologies, and so forth. So if there is one single lesson from the three case studies, it is that knowledge is always social in essence and therefore the management of knowledge is the management of social relations.

Appendix
On methodology

Methodology of the studies

The studies reported in this volume are based on a qualitative case study methodology (Eisenhardt, 1989; Becker, 1992; Stake, 1996) and more specifically on interview methodology. While interviewing is a very common research methodology in post-positivist research methods (P. Prasad, 2005), the epistemological status of interviews is still contested (Alvesson, 2003). For instance, Briggs (2003: 247) argues that it is assumed in 'Western ideologies of language' that interviews form the medium wherein 'referential content' is unproblematically transferred from one party to another. Similarly, Silverman (1993: 90) argues that a positivist perspective on interviews assumes that they 'give us access [to] "facts" about the world'. Such criticism has left social science researchers with a lingering doubt about what an interview is and what it can accomplish. Gubrium and Holstein (2003: 3) write: 'Interview roles are less clear than they once were ... Standardized representation has given way to representational invention, where the dividing line between fact and fiction is blurred to encourage richer understanding'. Those sceptical of interviews, for instance, the Czech writer Milan Kundera, argue that interviews are acts of symbolic violence where authenticity is claimed on the basis of loose conversations. Kundera says:

> [T]he interview as it is generally practiced has nothing to do with dialogue: (1) the interviewer asks questions of interest to him, of no interest to you; (2) of your responses, he uses only those that suit him; (3) he translates them into his own vocabulary, his own manners of thought.
>
> (Kundera, 1988: 133)

However, one may also regard interviews as systematic interactions wherein one may gain insights into the life world of others (Kvale, 1996). Such a pragmatic view of interviewing is taken in this study. Although one cannot safeguard interviews against the influence of undesirable effects such as moral storytelling, misunderstandings and a selective perception of discussed topics on the part of the interviewer, it nevertheless remains an opportunity for getting a glimpse of the everyday work-life of other individuals (see e.g., Terkel, 1972). In addition to the interviews, a variety of printed documents, homepages and informal documentation have been used in the case studies. In three of the studies, research findings have been reported back to the participating firms and practitioners have commented on the research results.

Companies were selected on the basis of a variety of reasons, for instance, their size and position in the industry, as in the coaching study which aimed at including a major and a middle-sized Swedish construction company to enable comparative analyses, to more personal contacts (as in the case of the two architecture firms). In addition, the author has used the industry interest organizations, such as Svenska Byggindustriers Utvecklingsfond (SBUF) and Byggmästareföreningen, to present ongoing and planned research. In general, Swedish industry is often accessible for academic researchers and there is an admirable ambition among managers and co-workers in the industry to contribute to academic research.

Data collection and data analysis

Chapter 2: The coaching study

The site management coaching research project was designed to comprise three distinct phases. In the first phase, the research phase, about twenty interviews were conducted with site managers, their closest superior manager and the foreman at the site. These interviewees worked in two construction companies: one major company and one medium-sized firm. On the basis of these interviews, six site managers were selected by their companies and asked to join the next phase, the coaching phase, wherein one professional coach, hired to take care of the assignment, met the six site managers regularly during a period of one year (June 2005 to June 2006). The six managers had 10–20 years experience from the industry and had worked as site managers for between ten years. Their experience came mostly from running large-scale construction projects such as housing projects and public buildings (e.g., schools and hospitals). For instance, during the coaching phase, the six site managers were in charge of a major shopping mall with a budget of more than 15 million euros, a major research laboratory, three major real estate projects and a health care unit. Two of the site managers also handled two or more projects at the same time. All the construction projects were new constructions. All these projects were prestigious projects for the two companies and the selected site managers were what at times are called 'heavyweight project leaders' in the project management literature. The coach had a background as a military officer and had extensive formal training in human resource management and leadership. During the last few years, after leaving his military position, the coach had worked with a major leadership programme widely used in Swedish industry and in the Swedish construction industry. One of the human resource managers in the two participating companies had previously worked with the coach and regarded him as a qualified and widely acclaimed leadership development consultant suitable for the task of being in charge of the coaching. During the coaching phase, the six site managers were interviewed three times: after a few months' coaching activities, halfway into the process, and at the end, at the time of the termination of the project; for a comprehensive overview of interviewing, see e.g., Holstein and Gubrium (2003). Two academic researchers not involved in the coaching practices conducted the interviews. The interview duration was about 1.5 hours and were tape-recorded and transcribed by one of the researchers. In addition, the coach was interviewed twice during the process, in the middle and the end, to give his account of how the coaching worked.

The coaching procedure was designed by the coach and derived from his formal training in the behavioural sciences and experience from working with organization change and human resource management development projects. The coachees were asked to bring issues for discussion to the meetings but the coach also brought a number of tools and methods to the joint meetings. Such tools and methods addressed communication, conflicts, meeting procedures and other topics he thought were of relevance for the practising site managers. The entire coaching phase was based on the site managers' own interests and will to develop their own skills and competencies as site managers. The academic researchers declared from the outset that their objective was to follow the process and its outcomes, and not to influence the content of the coaching. In addition, during the year-long coaching process, all of the six site managers and the coach met at a number of times to jointly reflect on their progress and what they had learned during the process. The site managers were then given the chance to share their experiences from the coaching with peers in the same position. The site managers also appreciated being given the chance to meet with and speak freely with fellow site managers, a group of colleagues they rarely had the time to see or speak to during their busy everyday work besides in a few formal meetings at the head office. During the coaching process, tentative research results were reported to a reference group, consisting of representatives of a major construction company, a smaller construction company, and two of the major Swedish trade unions representing managers and the blue-collar workers in the construction industry, respectively. The reference group delegates were asked to comment on their procedure and to provide suggestions on how coaching could be made a useful resource for improving the competitiveness of the construction industry.

Chapter 3: The architect study I

The case study of the Blue Architect Firm is based on six interviews with the chief architect (owner and chief executive officer; CEO), lead architect, engineer, three-dimension visualizer and project leader. The selection of interviewees was based on taking all members for one particular project, the national science centre, which the firm was involved in during 2001. A doctoral student and the author conducted the semi-structured interviews in the Fall and Winter 2003–4. The interview duration time was approximately one hour. The interviews were tape-recorded and transcribed by a professional transcriber.

Chapter 4: The architect study II

Twelve co-workers in two offices (the headquarters and a smaller office) were interviewed during the period April to May 2008 by two researchers. The sample included practising architects, managers (in all cases also with an architecture degree), a construction engineer and an interior decorator. The sample included both men and women, and included both co-workers with substantial organizational tenure (i.e., more than 25 years of work experience in the company) to more newly hired co-workers. Interviews lasted for about one hour and were tape-recorded. One of the researchers transcribed the interviews.

Chapter 5: The rock construction work study

The CEO and thirteen co-workers at ConCo were interviewed in May and June 2007. Interviews were semi-structured, tape-recorded and transcribed by the author. All interviews were conducted by the same researcher. Interviewees were selected by the CEO on the basis of the requests from the researcher. Interviewees included site managers and maintenance workers. No construction workers were interviewed but two of the site managers had worked their way up to a site manager's position. Some of the site managers had substantial organizational tenure, while others were more recent newcomers to the firm. Three of the interviewees had Masters degrees in engineering, while the rest were high-school graduates or had vocational training or some colleague education. All of the interviewees were men since all of the co-workers in the company, besides a few administrators, were male.

Data analysis in the four studies

The interview transcripts were coded (Huberman and Miles, 1994; Strauss and Corbin, 1998) by either of the two researchers participating in the data collection. Interview excerpts were structured into categories, which were further organized into a relevant series of events. That is, the categories were used to 'emplot' (White, 1987; Czarniawska, 1997; Patriotta, 2003) in to a sequence that would make sense for the readers. In two of the studies, research findings were reported back to the participating companies and elicited positive responses. Since all the interviews were conducted in Swedish, the interview excerpts had to be translated into English. Since there are obviously some nuances and 'tone' in the original utterances that may be 'lost in translation' (Hoffman, 1989), the translation of the interview excerpts aimed at capturing the sense and meaning of the original utterances. For instance, idiomatic expressions or commonplace sayings were translated into comparative English expressions.

Notes

Introduction

1 The degree of innovation in construction work is much debated and remains a contested terrain. Winch (2003), for instance, claims that the statistics are unfavourable for the construction industry in comparison to, for instance, the automotive industry, one of the benchmarks for construction. 'There is', Winch, (2003: 654) concludes, 'currently no reliable evidence that "construction", when compared to other industries properly considered as value systems has a lower (or higher) rate of innovation than its competitors'.

1 Occupational groups and professions, practices, institutions and knowledge in construction work

1 Another explanation for the relatively poor degree of innovation is proposed by Winch (1998), suggesting that innovation in construction takes place within what has been called the *complex product system* characterized by three properties: (1) 'many interconnected and customized elements organized in an hierarchical way'; (2) 'nonlinear and continuously emerging properties where small changes to one element of the system lead to larger changes elsewhere in the system'; and (3) 'a high degree of user involvement in the innovation process' (Winch, 1998: 269). In complex production systems, innovation engages several loosely coupled, heterogeneous social systems in a non-linear manner and the individual construction projects are in this view one particular social system among others. This perspective is also discussed by Gann (2000: 213) suggesting that 'Innovation in construction can ... only be properly understood through an examination of changes both on and off sites at sectoral and industrial boundaries. The analysis needs to focus on the construction system as a whole, including inputs and outputs and the relationship between the many different participants – from clients and designers, manufacturers and suppliers of materials, components and equipment to contractors and specialist engineering firms'. Innovation in the construction industry arguably should be examined on a systemic level and as the outcome from intricate technological, material, political and social collaborations over organizational boundaries.

Bibliography

Acker, Joan (2006) *Class Questions, Feminist Answers*, Lanham: Rowman & Littlefield.

Adler, Nancy (2006) The arts and leadership: now that we can do anything, what will we do?, *Academy of Management Learning and Education*, 5(4): 486–499.

Adler, Patricia A. (1985) *Wheeling and Dealing: An Ethnography of an Upper-level Drug Dealing and Smuggling Community*, New York: Columbia University Press.

Adler, Paul S. and Kwon, Soek-Woo. (2002) Social capital: prospects for a new concept, *Academy of Management Review*, 27(1): 17–40.

Adorno, Theodor W. (2000) *Introduction to Sociology*, Cambridge: Polity Press.

Agapiou, Andrew (2002) Perceptions of gender roles and attitudes toward work among male and female operatives in a Scottish construction industry, *Construction Economics and Management*, 20: 697–705.

Alvesson, Mats (1994) Talking in organizations: managing identity and impression in an advertising agency, *Organization Studies*, 15(4): 535–563.

Alvesson, Mats (2000) Social identity and the problem of loyalty in knowledge-intensive companies, *Journal of Management Studies*, 37(8): 1101–1123.

Alvesson, Mats (2001) Knowledge work: ambiguity, image and identity, *Human Relations*, 54(7): 863–886.

Alvesson, Mats (2003) Beyond neopositivists, romantics, and localists. a reflexive approach to interviews in organization research, *Academy of Management Review*, 28(1): 13–33.

Amin, Ash and Cohendet, Patrick (2004) Architecture of Knowledge. *Firms, Capabilities, and Communities*, Oxford and New York: Oxford University Press.

Andreu, Isabel Carmona and Oreszczyn, Tadj (2004) Architects need environmental feedback, *Building Research and Information*, 32(4): 313–328.

Ankrah, N.A. and Langford, D.A. (2005) Architects and contractors: a comparative study of organizational cultures, *Construction Management and Economics*, 23: 595–607.

Applebaum, Herbert A. (1981) *Royal Blue: The Culture of Construction Workers*, New York: Holt, Rinehart, Winston.

Argyris, Chris and Schön, Donald A. (1978) *Organizational Learning: A Theory of Action Perspective*, Reading, MA: Addison-Wesley.

Arnaud, Gilles (2003) A coach or a couch? A Lacanian perspective on executive coaching and consulting, *Human Relations*, 56(9): 1131–1154.

Attewell, Paul (1990) What is a skill?, *Work and Occupations*, 17(4): 422–448.

Austin, John L. (1962) *How To Do Things With Words*, Oxford: Oxford University Press.

Bachelard, G. (1934/1984) *The New Scientific Spirit*, Boston: Beacon Press.

Bain, J. (1968) *Industrial Organization*, New York: Wiley.

Bakken, Tore and Hernes, Tor, Eds. (2003) *Autopoetic Organization Theory: Drawing on Niklas Luhmann's Social Systems Perspective*, Oslo: Abstrakt; Malmö: Liber; Copenhagen: Copenhagen Business School Press.

Balkundi, Prasad and Kilduff, Martin (2005) The ties that lead: a social network approach to leadership, *Leadership Quarterly*, 16: 941–961.

Ball, Kristie and Carter, Chris (2002) The charismatic gaze: everyday leadership practices of the 'new' manager, *Management Decision*, 40(6): 552–565.

Barley, Stephen R. (1986) Technology as an occasion of structuring: Evidence from observations of CT scanners and the social order of radiology departments, *Administrative Science Quarterly*, 31: 78–108.

Barley, Stephen R. (1990) The alignment of technology and structure through roles and networks, *Administrative Science Quarterly*, 35: 61–103.

Barley, Stephen R. and Kunda, Gideon (2004) *Gurus, warm bodies and hired guns: Itinerant experts in the knowledge economy*, Princeton: Princeton University Press.

Barley, Stephen R. and Kunda, Gideon (2006) Contracting: a new form of professional practice, *Academy of Management Perspectives*, 20(1): 45–66.

Barley, Stephen R. and Tolbert, Pamela (1997) Institutionalization and structuration. Studying the links between action and institution, *Organization Studies*, 18(1): 93–117.

Barney, J. (1991) Firm resources and sustained competitive advantage, *Journal of Management*, 17: 99–120.

Barney, Jay B. (2001) Is the resource-based 'view' a useful perspective for strategic management research? Yes, *Academy of Management Review*, 26(1): 41–56.

Bartel, Caroline A. and Garud, Raghu (2003) Narrative knowledge in action: adaptive abduction as a mechanism for knowledge creation and exchange in organizations, in Easterby-Smith, Mark and Lyles, Marjorie A (2003) *Handbook of Organization Learning and Knowledge Management*, Oxford and Malden: Blackwell, pp. 324–342.

Bateson, Gregory (1979) *Mind and Nature: A Necessary Unit*, New York: E.P. Dutton.

Bechky, Beth A. (2003a) Object lessons: workplace artifacts as representations of occupational jurisdiction, *American Journal of Sociology*, 109(3): 720–752.

Bechky, Beth A. (2003b) Object lessons: workplace artifacts as representations of occupational jurisdiction, *American Journal of Sociology*, 109(3): 720–752.

Becker, Howard S. (1992) Cases, causes, conjunctures, stories, and imagery, in Ragin, Charles C. and Becker, Howard S., Eds. (1992) *What is a Case?: Reexploring the Foundations of Social Inquiry*, Cambridge. Cambridge University Press.

Becker, Howard S., Geer, Blanchie, Hughes, Everett C. and Strauss, Anselm L. (1961) *Boys in White: Student Culture in Medical School*, Chicago: The University of Chicago Press.

Becker, Mattias C. (2004) Organizational routines; a review of the literature, *Industrial and Corporate Change*, 13(4): 643–677.

Bell, Daniel (1973) *The Coming Post-Industrial Society*, New York: Basic Books.

Beller, Jonathan (2006) *The Cinematic Mode of Production: Attention Economy and the Society of the Spectacle*, Duke Hanover: Dartmouth College Press.

Belova, Olga (2006) The event of seeing: a phenomenological perspective on visual sense-making, *Culture and Organization*, 12(2): 93–107.

Bendix, Reinhard (1971) Bureaucracy, in Bendix, Reinhard and Roth, Guenther Eds. (1871) *Scholarship and Partisanship: Essays on Max Weber*, Berkeley, Los Angeles and London: University of California Press, pp. 129–155.

Benjamin, Walter (1999) *The Arcades Project*, trans. by Eiland, Howard and McLaughlin, Kevin, Cambridge: The Belknap Press.

Berglas, Steven (2002) Dangers of executive coaching, *Harvard Business Review*, June, pp. 86–92.

Berle, Adolf A. and Means, Gardiner C. (1934/1991) *The Modern Corporation and Private Property*, New Brunswick: Transaction Publishers.

Blau, Judith R. (1984) *Architects and Firms: A Sociological Perspective on Architectural Practice*, Cambridge: MIT Press.

Blau, Judith R. and McKinley, William (1979) Ideas, complexity and innovation, *Administrative Science Quarterly*, 24: 200–219.

Boden, Deidre (1994) *The Business of Talk: Organizations in Action*, Cambridge: Polity Press.

Böhme, Gernot (2003) Contribution to the critique of the aesthetic economy, *Thesis Eleven*, 73: 72–82.

Boisot, Max H. (1998) *Knowledge Assets: Securing Competitive Advantage in the Information Economy*, Oxford: Oxford University Press.

Boje, David M. (1991) The storytelling organization: a study of story performance in an office supply firm, *Administrative Science Quarterly*, 36: 106–126.

Boland, Richard J., Lyytinen, Kalle and Yoo, Youngjin (2007) Wakes of innovation in project networks: The case of digital 3-D representation in architecture, engineering, and construction, *Organization Science*, 18(4): 631–647.

Bontis, Nick, Crossan, Mary M. and Hulland, John (2002) Managing an organization learning system by aligning stocks and flows, *Journal of Management Studies*, 39(4): 438–469.

Bourdieu, Pierre (2004) *The Science of Science and Reflexivity*, Chicago and London: The University Of Chicago Press.

Bowen, Paul, Pearl, Robert and Akintoye, Akintola (2007) Professional ethics in the South African construction industry, *Building Research and Information*, 35(2): 189–205.

Bowker, Geoffrey C. and Leight Star, Susan (1999) *Sorting Things Out: Classification and its Consequences*, Cambridge and London: MIT Press.

Brand, Stewart (1994) *How Buildings Learn: What Happens After They're Built*, London: Viking.

Braudel, Fernand (1992) *The Wheels of Commerce: Civilization and Capitalism 15th–18th century*, Vol. 2, Berkeley and Los Angeles: The University of California Press.

Bresnen, Mike, Edelman, Linda, Newell, Sue, Scarbrough, Harry and Swan, Jacky (2005) Exploring social capital in the construction firm, *Building Research and Information*, 33(3): 235–244.

Bresnen, Mike, Goussevskaia, Anna and Swan, Jacky (2004) Embedding new management knowledge in project-based organizations, *Organization Studies*, 25(9): 1535–1555.

Briggs, Charles L. (2003) Interviewing, power/knowledge, and social inequality, in Holstein, James A. and Gubrium, Jaber F., Eds. (2003) *Inside Interviewing*.

New Lenses, New Concerns, London: Thousand Oaks and New Delhi: Sage, pp. 495–506.

Brunsson, Nils (1985) *The Irrational Organization: Irrationality as a Basis for Organizational Action and Change,* Wiley, New York.

Bryson, Norman (1988) The gaze in the expanded field, in Foster, Hal, Ed. (1988) *Vision and Visuality,* New York: The New Press, pp. 86–113.

Bucciarelli, Louis, L. (1994) *Designing Engineers,* Cambridge and London: MIT Press.

Buckle, Pamela and Thomas, Janice (2003) Deconstructing project management: a gender analysis of project management guidelines, *International Journal of Project Management,* 21: 433–441.

Burawoy, Michael (1979) *Manufacturing Consent: Changes in the Labour Process under Monopoly Capitalism,* Chicago: University of Chicago Press.

Burdett, John O. (1998) Forty things every manager should know about coaching, *Journal of Management Development,* 17(2): 142–152.

Burns, Alfred (1989) *The Power of the Written Word: The Role of Literacy in the History of Western Civilization,* New York: Peter Lang.

Burt, Ronald S. (1997) The contingent value of social capital, *Administrative Science Quarterly,* 42: 339–365.

Butler, Arthur G. (1973) Project management: a study in organizational conflict, *Academy of Management Journal,* 16(1): 84–101.

Callon, Michel (2002) Writing and (re)writing devices as tools for managing complexity, in Law, John and Mol, Annemarie, Eds. (2002) *Complexities: Social Studies of Knowledge Practices,* Durham and London: Duke University Press.

Canguilhem, George (1989) *A Vital Rationalist: Selected Writings from George Canguilhem,* New York: Zone Books.

Carlsson, Sune (1951) *Executive Behaviour: A Study of the Work Load and the Working Methods of Managing Directors,* Stockholm: Strömbergs.

Carrillo, Patricia (2004) Managing knowledge: lessons from the oil and gas sector, *Construction Management and Economics,* 22: 631–642.

Carrillo, Patricia, Robinson, Herbert, Al-Ghassani, Ahmed and Anumba, Chimay (2004) Knowledge management in UK construction. Strategies, resources and barriers, *Project Management Journal,* April, 46–56.

Castoriadis, Cornelius (1997) *World in Fragments,* Stanford: Stanford University Press.

Chia, Robert and Holt, Robin (2008) On management knowledge, *Management Learning,* 39(2): 141–158.

Cicmil, Svetlana and Hodgson, Damian (2006) Making projects critical: An introduction, in Hodgson, Damian and Cicmil, Svetlana, Eds. (2006) *Making Projects Critical,* Basingstoke and New York: Palgrave, pp. 1–25.

Clarke, Adele E. and Fujimura, Joan H. (1992) What tools? Which jobs? Why right?, in Clarke, Adele E. and Fujimura, Joan H., Eds. (1992) The Right Tools for the Job. At Work in Twentieth-Century Life Sciences, Princeton: Princeton University Press, pp. 3–45.

Clarke, Adele E., Mamo, Laura, Fishman, Jennifer R., Shim, Janet K. and Fosket, Jennifer Ruth (2003) Biomedicalization. Technoscientific transformations of health, illness, and U.S. biomedicine, *American Sociological Review,* 68: 161–194.

Clegg, Stewart R., Rhodes, Carl and Kornberger, Martin (2007) Desperately seeking legitimacy: organizational identity and emerging industries, *Organization Studies*, 28(4): 495–513.

Cohen, Dan (2007) Enhancing social capital for knowledge effectiveness, in Ichijo, Kazuo, Nanaka, Ikujiro, Eds. (2007) *Knowledge Creation and Management: New Challenges for Managers*, Oxford and New York; Oxford University Press, pp. 240–253.

Cohen, Michael D., March, James G. and Olsen, Johan P. (1972) A garbage can model of organizational choice, *Administrative Science Quarterly*, 17: 1–25.

Coleman, James S. (1988) Social capital in the creation of human capital, *American Journal of Sociology*, 94: S95–121.

Collins, Randall (1979) *The Credential Society*, New York: Academic Press.

Conner, K. (1991) A historical comparison of resource-based theory and five schools of thought within industrial organization economics: do we have a new theory of the firm?, *Strategic Management Journal*, 17: 121–154.

Construction Statistics Annual Report (2006) London: Department of Trade and Industry.

Cooley, Charles Horton (1902/1964) *Human Nature and the Social Order*, New York: Schocken Books.

Crary, Jonathan (1990) *Techniques of the Observer: On Vision and Modernity in the Nineteenth Century*, Cambridge and London: MIT Press.

Crary, Jonathan (1995) Unbinding vision: Manet and the attentive observer in the late Nineteenth century, in Carney, Leo and Schwartz, Vanessa R., Eds. (1995) *Cinema and the Invention of Modern Life*, Berkeley: University of California Press, pp. 46–71.

Crary, Jonathan (1999) *Suspensions of Perception: Attention, Spectacle, and Modern Culture*, Cambridge and London: MIT Press.

Croce, Benedetto (1990/1992) *The Aesthetics as the Science of Expression and of the Linguistic in General*, Cambridge: Cambridge University Press.

Cuff, Dana (1991) *Architecture: The Story of Practice*, Cambridge: MIT Press.

Currie, Graeme and Brown, Andrew D. (2003) A narratological approach to understanding processes of organizing in a UK hospital, *Human Relations*, 56(5): 563–586.

Czarniawska, Barbara (1997) *Narrating the Organization: Dramas of Institutional Identity*, Chicago and London: The University of Chicago Press.

Czarniawska, Barbara (2004) *Narratives in Social Science Research*, London: Thousand Oaks and New Delhi: Sage.

Czarniawska, Barbara (2005) On Gorgon sisters: organizational action in the face of paradox, in Seidl, David and Becker, Kai Helge, Eds. (2005) *Niklas Luhmann and Organisation Studies*, Copenhagen: Copenhagen Business School Press, pp. 127–142.

Dadfar, Hossein and Gustafsson, Peter (1992) Competition by effective management of cultural diversity: the case of international construction projects, *International Studies of Management and Organizations*, 22(4): 81–92.

Dahrendorf, Ralf (1959) *Class and Class Conflict in Industrial Society*. London: Routledge & Kegan Paul.

Dainty, Andrew R.J., Bagilhole, Barbara M., Ansari, K.H. and Jackson, J. (2004) Creating equality in the construction industry. An agenda for change for women and ethnic minorities, *Journal of Construction Research*, 5(1): 75–86.

Dalton, Melville (1959) *Men Who Manage: Fusion of Feeling and Theory in Administration*, New York: Wiley.

Daston, Loraine and Galison, Peter (2007) *Objectivity*, New York: Zone Books.

Daston, Lorraine (2008) On scientific observation, *Isis*, 99: 97–110.

Davidson, Marilyn J. and Sutherland, Valerie J. (1992) Stress and construction site managers: Issues for Europe 1992, *Employee Relations*, 14(2): 25–38.

DeLanda, Manuel (2006) *A New Philosophy of Society: Assemblage Theory and Social Complexity*, London and New York: Continuum.

de Monthoux, Pierre Guillet (2004) *The Art Firm: Aesthetic Management and Metaphysical Marketing from Wagner to Wilson*, Stanford: Stanford University Press.

Delmestri, Guiseppe and Walgenbach, Peter (2005) Mastering techniques or brokering knowledge? Middle managers in Germany, Great Britain and Italy, *Organization Studies*, 26(2): 197–220.

Dent, Mike (2003) Managing doctors and saving a hospital: irony, rhetoric and actor networks, *Organization*, 10(1): 107–127.

de Saussure, Ferdinand, (1959) *Course in General Linguistics*, London: Peter Owen.

Denzin, Norman K. (1995) *The Cinematic Society: The Voyeur's Gaze*, London: Thousand Oaks and New Delhi: Sage.

Dierkes, Meinolf, Berthon, Ariane, Child, John and Nonaka, Ikujiro, Eds. (2001) *Handbook of Organizational Learning and Knowledge*, Oxford: Oxford University Press.

DiMaggio, Paul and Powell, Walter W. (1983) The iron cage revisited: Institutional isomorphism and collective rationality in organizational fields, *American Sociological Review*, 48(2): 147–160.

DiMaggio, Paul J. and Powell, Walter W. (1991) *The New Institutionalism in Organizational Analysis*, Chicago: University of Chicago Press.

Djerbarni, R. (1996) The impact of stress in site management effectiveness, *Construction Management and Economics*, 14: 281–293.

Donnellon, Anne (1996) *Team Talk: Listening Between the Lines to Improve Team Performance*, Boston: Harvard Business School Press.

Dopson, Sue and Stewart, Rosemary (1990) What is happening to middle management?, *British Journal of Management*, 1: 3–16.

Dorée, André G. and Holmen, Elizabeth (2004) Achieving the unlikely: innovation in loosely coupled systems, *Construction Economics and Management*, 22: 827–838.

Dougherty, Deborah (2007) Trapped in the 20th century? Why models of organizational learning, knowledge and capabilities do not fit bio-pharmaceuticals, and what to do about that, *Management Learning*, 38(3): 265–270.

Dougherty, Deborah and Takacs, C. Helen (2004) Team play: heedful interrelating as the boundary for innovation, *Long Range Planning*, 37(6): 569–590.

Douglas, Mary (1986) *How Institutions Think*, London: Routledge & Kegan Paul.

Drejer, Ina and Vinding, Anker Lund (2006) Organization, 'anchoring' of knowledge, and innovative activity in construction, *Building Research and Information*, 24: 921–931.

Druskat, Vanessa Urch and Pescosolido, Anthony (2002) The content of effective teamwork mental models in self-managing teams: ownership, learning and heedful interrelating, *Human Relations*, 55(3): 283–314.

Dubois, Anna and Gadde, Lars-Erik (2002) The construction industry as a loosely coupled system: implications for productivity and innovation, *Construction Management and Economics*, 20: 621–631.

Eagleton, Terry (1990) *The Ideology of the Aesthetics*, Oxford and Cambridge: Blackwell.

Easterby-Smith, Mark and Lyles, Marjorie A. (2003) *Handbook of Organization Learning and Knowledge Management*, Oxford and Malden: Blackwell.

Eccles, Robert G. (1981) Bureaucratic versus craft administration: the relationship of market structure to the construction firm, *Administrative Science Quarterly*, 26(3): 449–469.

Edenius, Mats and Yakhlef, Ali (2007) Space, vision, and organizational learning; the interplay of incorporating and inscribing practices, *Management Learning*, 38(2): 193–210.

Edmonson, Amy C. (2003) Speaking up in the operating room: how team leaders promote learning in interdisciplinary action teams, *Journal of Management Studies*, 40(6): 1419–1452.

Eisenhardt, Kathleen N. (1989) Building theories from case study research, *Academy of Management Review*, 14(4): 532–550.

Empson, Laura (2001) Introduction: knowledge management in professional service firms, *Human Relations*, 54(7): 811–817.

English, Jane (2002) Managing cultural differences to improve industry efficiency, *Building Research and Information*, 30(3): 196–204.

Entwhistle, Joanna and Racamora, Agnès (2006) The field of fashion materialized: a study of London Fashion Week, *Sociology*, 40(4): 735–751.

Etzioni, Amitai (1964) *Modern Organizations*, Englewood Cliffs: Prentice-Hall.

Ewenstein, Boris and Whyte, Jennifer (2007a) Beyond words: aesthetic knowledge and knowing in organizations, *Organization Studies*, 28(5): 689–708.

Ewenstein, Boris and Whyte, Jennifer K. (2007b) Visual representations as 'artifacts of knowing', *Building Research and Information*, 35(1): 81–89.

Faulkner, Wendy (2007) 'Nuts and bolts and people': gender-troubled engineering identities, *Social Studies of Science*, 37(3): 331–356.

Feldman, Martha S. (2000) Organization routines as a source of continuous change, *Organization Science*, 11(6): 611–629.

Feldman, Martha S. and Pentland, Brian T. (2003) Reconceptualizing organization routines as a source of flexibility and change, *Administrative Science Quarterly*, 48: 94–118.

Feldman, Martha and Pentland, Brian (2005) Organizational routines and the macro-actors, in Czarniawska, Barbara and Hernes, Tor (2005) *Actor-Network Theory and Organizing*, Malmö: Liber and Copenhagen: Copenhagen Business School Press, pp. 91–111.

Feldman, Martha S. and Rafaeli, Anat (2002) Organizational routines as sources of connections and understandings, *Journal of Management Studies*, 39(3): 309–331.

Fine, Gary Alan (1996) *Kitchens: The Culture of Restaurant Work*, Berkeley, Los Angeles and London: University of California Press.

Fitzgerald, Paula and Ellen, Pam Scholder (1999) Scents in the marketplace: explaining a fraction of olfaction, *Journal of Retailing*, 75(2): 243–262.

Fligstein, Neil (1990) *The Transformation of Corporate Control*, Cambridge and London: Harvard University Press.

Florida, Richard (2002) *The Rise of the Creative Class*, New York. Basic Books.

Floyd, Steven W. and Woolridge, Bill. (1997) Middle management's strategic influence and organizational performance, *Journal of Management Studies*, 34(3): 465–485.

Flynn, Francis J. and Staw, Barry M. (2003) Lend me your wallet: the effect of charismatic leadership on external support or an organization, *Strategic Management Journal*, 25: 309–330.

Foucault, M. (1973) *The Birth of the Clinic*, London: Routledge.

Foucault, M. (1977) *Discipline and Punish*, New York: Pantheon.

Fowler, Bridget and Wilson, Fiona (2004) Women architects and their discontents, *Sociology*, 38(1): 101–119.

Frank, David John and Meyer, John W. (2007) University expansion and the knowledge society, *Theoretical Sociology*, 36: 287–311.

Fraser, C. (2000) The influence of personal characteristics on effectiveness of construction site managers, *Construction Management and Economics*, 18: 29–36.

Freidson, Eliot (1986) *Professional Powers: A Study of the Institutionalization of Formal Knowledge*, Chicago and London: The University of Chicago Press.

Fuchs, Stephan (1992) *The Professional Quest for Truth: A Social Theory of Science and Knowledge*, Albany: State University of New York Press.

Fujimura, Joan H. (1996) *Crafting Science: A Sociohistory of the Quest for the Genetics of Cancer*, Cambridge, MA: Harvard University Press.

Gabriel, Yannis (2000) *Storytelling in Organizations: Facts, Fictions, and Fantasies*, Oxford: Oxford University Press.

Gann, David M. (2000) *Building Innovation: Complex Constructs in a Changing World*, London: Thomas Telford.

Gantt, Henry L. (1913/1919) *Work, wages, and profits*, 2nd ed., New York: The Engineering Magazine Co.

Garfinkel, Harold (1967) *Studies in Ethnomethodology*, Prentice-Hall, Englewood Cliffs.

Gherardi, Silvia (2001) From organizational learning to practice-based knowing, *Human Relations*, 54(1): 131–139.

Gherardi, Silvia (2006) *Organizational Knowledge: The Texture of Workplace Learning*, Cambridge and Malden: Blackwell.

Gherardi, Silvia and Nicolini, Davide (2001) The sociological foundations of organizational learning, in Dierkes, Meinolf, Berthon, Ariane, Child, John and Nonaka, Ikujiro, Eds. (2001) *Handbook of Organizational Learning and Knowledge*, Oxford: Oxford University Press.

Gherardi, Silvia and Nicolini, Davide (2002) Learning the trade: a culture of safety in practice, *Organization*, 9(2): 191–223.

Gibson, Lisanne (2002) Managing the people: art programs in the American depression, *Journal of Arts Management, Law, and Society*, 31(4): 279–291.

Giddens, Anthony (1984) *The Constitution of Society*, Chicago: The University of Chicago Press.

Giddens, Anthony (1990) *The Consequences of Modernity*, Cambridge: Polity Press.

Gieryn, Thomas F. (1983) Boundary-work and the demarcation of science from non-science: strains and interest in professional ideologies of scientists, *American Sociological Review*, 48(6): 781–795.

Goffman, E. (1961) *Asylums*, London: Penguin.

Goodwin, Charles (1994) Professional vision, *American Anthropologist*, 96(3): 606–633.

Goodwin, Charles (1995) Seeing in depth, *Social Studies of Science*, 25: 237–274.

Goodwin, Charles (2001) Practices of seeing visual analysis: an ethnomethdological approach, in Van Leeuwen, Theo and Jewitt, Carey, Eds. (2001) *Handbook of Visual Analysis*, London, Thousand Oaks and New Delhi: Sage, pp. 157–182.

Goody, Jack (1986) *The Logic of Writing and the Organization of Society*, Cambridge: Cambridge University Press.

Gourlay, Stephen (2006) Conceptualizing knowledge creation: a critique of Nonaka's theory, *Journal of Management Studies*, 43(7): 1415–1436.

Gramsci, Antonio (1971) *Selection from Prison Notebooks*, New York: International Publishers.

Granovetter, Mark S. (1973) The strength of weak ties, *American Journal of Sociology*, 78(6): 1360–1380.

Gray, David E. (2006) Executive coaching: towards a dynamic alliance of psychotherapy and transformative learning processes, *Management Learning*, 37(4): 475–497.

Green, Stuart (2006) The management of projects in the construction industry: context, discourse and self-identity, in Hodgson, Damian and Cicmil, Svetlana, Eds. (2006) *Making Projects Critical*, Basingstoke and New York: Palgrave, pp. 207–231.

Grosz, Elizabeth (2001) *Architecture from the Outside: Essays on Virtual and Real Spaces*, Cambridge: MIT Press.

Gubrium, Jaber F, and Holstein, James A., Eds. (2003) Postmodern sensibilities, in Gubrium, Jaber F. and Holstein, James A., Eds. (2003) *Postmodern Interviewing*, London, Thousand Oaks and New Delhi: Sage, pp. 3–16.

Guillén, Mauro F. (1994) *Models of Management: Work, Authority, and Organization in a Comparative Perspective*, Chicago and London: The University of Chicago Press.

Guillén, Mauro F. (1997) Scientific management's lost aesthetic: architecture, organization, and the Taylorized beauty of the mechanical, *Administrative Science Quarterly*, 42: 682–715.

Guler, Isin, Guillén, Mauro F. and Macpherson, John Muir (2002) Global competition, institutions, and the diffusion of organizational practices: the international spread of ISO 9000 quality certificates, *Administrative Science Quarterly*, 47: 207–232.

Gutman, Robert (1988) *Architectural Practice: A Critical View*, Princeton: Princeton Architectural Press.

Guve, Bertil Gonzàlez (2007) Aesthetics of financial judgments: on risk capitalists' confidence, in Guillet de Monthoux, Pierre, Gustafsson, Claes and Sjöstrand, Sven-Erik, Eds. (2007) *Aesthetic Leadership: Managing Fields of Flow in Art and Business*, Basingstoke: Palgrave Macmillan, pp. 128–140.

Hackman, J. Richard and Wageman, Ruth (2005) A theory of team coaching, *Academy of Management Review*, 30(2): 269–287.

Hall, Douglas T, Otazo, Karen I. and Hollenbeck, George P. (1999) What really happens in executive coaching, *Organization Dynamics*, 27(3): 39–53.

Hallyn, Fernand (1987/1990) *The Poetic Structure of the World: Copernicus and Kepler*, New York: Zone Books.

Hancock, Philip (2005) Uncovering the semiotics in organizational aesthetics, *Organization*, 12(1): 29–50.

Hancock, Philip and Tyler, Melissa (2007) Un/doing gender and the aesthetic of organizational performance, *Gender, Work and Organization*, 14(6): 512–532.

Haraway, D. (1997). *Modest Witness @ Second Millennium FemaleMan_ Meets_Onco Mouse: Feminism and Technoscience*. New York and London: Routledge.

Haraway, Donna J. (2000) *How Like a Leaf: An Interview with Thyrza Nichols Goodeve*, New York and London: Routledge.

Hargadon, Andrew B. and Douglas, Yellowlees (2001) When innovations meet institutions: Edison and the design of the electric light, *Administrative Science Quarterly*, 46: 476–501.

Harré, Rom (2002) Material objects in social worlds, *Theory, Culture and Society*, 19(5/6): 23–33.

Harty, Chris (2005) Innovation in construction: a sociology of technology approach, *Building Research and Information*, 33(6): 512–522.

Hassard, John and Parker, Martin (1993) *Postmodernism and Organizations*, London: Thousand Oaks and New Delhi: Sage.

Hasselbladh, Hans and Kallinkos, Jannis (2000) The project of rationalization: a critique and reappraisal of neo-institutionalism in organization studies, *Organization Studies*, 21(4): 697–720.

Hassoun, Jean-Pierre (2005) Emotions on the trading floor; social and symbolic expressions, in Knorr Cetina, Karin and Preda, Alex, Eds. (2005) *The Sociology of Financial Markets*, Oxford and New York: Oxford University Press, pp. 102–120.

Heervagen, Judith H., Kampschroer, Kevin, Powell, Kevin M. and Loftness, Vivian (2004) Collaborative knowledge environments, *Building Research and Information*, 32(6): 520–528.

Henderson, Kathryn (1999) *On Line and on Paper: Visual Representations, Visual Culture, and Computer Graphics in Design Engineering*, Cambridge and London: MIT Press.

Hernes, Tor and Bakken, Tore (2004) Implications of self-reference: Niklas Luhmann's autopoesis and organization theory, *Organization Studies*, 24(9): 1511–1535.

Herrstein Smith, Barbara (2005) *Scandalous Knowledge: Science, Truth and the Human*, Durham and London: Duke University Press.

Hillebrandt, Patricia M. (2000) *Economic Theory and Construction Industry*, 3rd edn., Basingstoke: Macmillan.

Hobday, Mike (2000) The project-based organisation. An ideal for managing complex products and systems?, *Research Policy*, 29: 871–893.

Hodgson, Damian (2002) Disciplining the professional: The case of project management, *Journal of Management Studies*, 39(6): 803–821.

Hodgson, Damian E. (2004) Project work. The legacy of bureaucratic control in the post-bureaucratic organization, *Organization*, 11(1): 81–100.

Hodgson, Damian (2005) Putting on a professional performance. Performativity, subversion and project management, *Organization*, 12(1): 51–68.

Hoffman, Eva (1989) *Lost in Translation: A Life in a New Language*, London: Minerva.

Hodgson, Damian and Cicmil, Svetlana (2007) The politics of standards in modern management: Making 'the project' a reality, *Journal of Management Studies*, 44(3): 431–450.

Holstein, James A. and Gubrium, Jaber F., Eds. (2003) *Inside Interviewing. New Lenses, New Concerns,* London: Thousand Oaks and New Delhi: Sage.

Hoopes, David G. and Postrel, Steven (1999). Shared knowledge, 'glitches', and product development performance. *Strategic Management Journal,* 20: 837–865.

Huemer, Lars and Östergren, Katarina (2000) Strategic change and organization learning in two 'Swedish' construction firms, *Construction Management and Economics,* 18: 635–642.

Hughes, Everett Cherrington (1958) *Men and Their Work,* Glencoe, IL: The Free Press.

Huy, Quy Nguyen (2002) Emotional balancing of organizational continuity and radical change: the contribution of middle managers, *Administrative Science Quarterly,* 47: 31–69.

Illich, Ivan (1977) *Disabling Professions,* London: Marion Boyars.

Ingold, Timothy (2000) *The perception of the environment,* London and New York: Routledge.

Inkpen, Andrew C. and Tsang, Eric W.K. (2005) Social capital, networks, and knowledge transfer, *Academy of Management Review,* 30(1): 146–165.

Ivory, Chris (2004) Client, user, and architect interactions in construction: implications for analyzing innovative outcomes from user–producer interactions in projects, *Technology Analysis and Strategic Management,* 16(4): 495–508.

Jackall, Robert (1988) *Moral Mazes: The World of Corporate Managers,* Oxford and New York: Oxford University Press.

Janis, Irving L. (1982) *Groupthink: Psychological Studies of Policy Decisions and Fiascoes,* 2nd edn, Boston: Houghton Mifflin.

Jermier, John M., Slocum, John W. Jr., Fry, Louis W. and Gaines, Jeannie (1992) Organizational subcultures in a soft bureaucracy: resistance behind the myth and façade of an official culture, *Organization Science,* 2(2): 170–194.

Johnson, Ericka (2007) Surgical simulations and simulated surgeons: reconstituting medical practices and practitioners in simulations, *Social Studies of Science,* 37: 585–608.

Jordanova, Ludmilla (1989) *Sexual Visions: Images of Gender in Science and Medicine Between the Eighteenth and Twentieth Centuries,* London: Harvester Wheatsheaf.

Judge, William Q. and Cowell, Jeffrey (1997) The brave new world of executive coaching, *Business Horizons,* July–August, pp. 71–77.

Julier, Guy (2000) *The Culture of Design,* London: Thousand Oaks and New Delhi: Sage.

Kadefors, Anna (1995) Institutions in building projects: implications for flexibility and change, *Scandinavian Journal of Management,* 11(4): 395–408.

Kallinikos, Jannis (2003) Work, human agency and organizational forms: an anatomy of fragmentation, *Organization Studies,* 24(4): 595–618.

Kamara, J.M., Augenbroe, G., Carrillo, P.M. (2002) Knowledge management in the architecture, engineering and construction industry, *Construction Innovation,* 2: 53–67.

Kampa-Kokesch, Sheila and Anderson, Mary Z. (2001) Executive coaching: a comprehensive review of the literature, *Consulting Psychology Journal: Practice and Research,* 53(4): 205–228.

Kärreman, Dan and Alvesson, Mats (2004) Cages in tandem: management control, social identity, and identification in a knowledge-intensive firm, *Organization*, 11(1): 149–175.

Kasson, John F. (1976) *Civilizing The Machine: Technology and Republican Values in America, 1776–1900*, New York: Grossman.

Kazi, Abdul Samad, Ed. (2005) *Knowledge Management in the Construction Industry: A Socio-Technical Perspective*, Hershey, PA: Idea Group.

Keeling, Ralph (2000) *Project Management*, London: Macmillan.

Knorr Cetina, Karin D. (1981) *The Manufacture of Knowledge: An Essay on the Constructivist and Contextual Nature of Science*, Oxford: Pergamon Press.

Knorr Cetina, Karin D. (1983) The ethnographic study of scientific work: towards a constructivist interpretation of science, in Knorr Cetina, Karin D. and Mulkay, Michael, Eds. (1983) *Science Observed: Perspectives on the Social Study of Science*, London, Beverly Hills and New Delhi: Sage, pp. 115–140.

Knorr Cetina, Karin (1997) Sociality with objects: social relations in postsocial societies, *Theory, Culture and Society*, 14(4): 1–30.

Koolhaas, Rem (1978) *Delirious New York*, New York: The Monticelli Press.

Kotter, John P. (1982) What effective general managers really do?, *Harvard Business Review*, Nov/Dec, 60(6): 156–168.

Koyré, Alexandre (1968/1992) *Metaphysics and Measurement*, Reading: Gordon and Breach Science Publishers.

Kundera, Milan (1988) *The Art of the Novel*, trans. by Asher, Linda, New York: Grove Press.

Kurland, Nancy B. and Pelled, Lisa Hope (2000) Passing the word: toward a model of gossip and power in the workplace, *Academy of Management Review*, 25(2): 428–438.

Kvale, Steinar (1996) *InterViewing*, London: Sage.

Kwinter, Sanford (2001) *Architecture of Time: Towards a Theory of the Event in Modernist Culture*, Cambridge and New York: MIT Press.

Lacan, Jacques (1998) *The Four Fundamental Concepts of Psychoanalysis: The Seminars of Jacques Lacan, Book XI*, trans. by Sheridan, Alan, New York and London: W.W. Norton.

Lamont, Michèle and Molnár, Virág (2002) The study of boundaries in social sciences, *Annual Review of Sociology*, 28: 167–195.

Lanzara, Giovan Francesco and Patriotta, Gerardo (2001) Technology and the courtroom: an inquiry into knowledge making in organizations, *Journal of Management Studies*, 38(7): 943–971.

Lanzara, Giovan Francesco and Patriotta, Gerardo (2007) The institutionalization of knowledge in an automotive factory: templates, inscriptions, and the problem of durability, *Organization Studies*, 28(5): 635–660.

Larson, Magali Sarafatti (1977) *The Rise of Professionalism: A Sociological Analysis*, Berkeley, Los Angeles and London: University of California Press.

Latour, B. (1987) *Science in Action*, Cambridge: Harvard University Press.

Latour, Bruno (1988) *The Pasteurization of France*, trans. by Sheridan, Alan and Law, John, Cambridge and London: Harvard University Press.

Latour, B. and Woolgar, S. (1979) *Laboratory Life: The Construction of Scientific Facts*, New Jersey: Princeton University Press.

Law, John (2002) Objects and spaces, *Theory, Culture and Society*, 19(5/6): 91–105.

Law, John and Whittaker, John (1988) On the art of representation: notes on the politics of visualization, in Fyfe Gordon, and Law, John, Eds. (1988) *Picturing Power: Visual Depiction and Social Relations*, London and New York: Routledge, pp. 160–183.

Le Corbusier (1946) *Towards a New Architecture*, trans. by Etchell, Frederick, London: The Architectural Press.

Le Goff, Jacques (1993) *Intellectuals in the Middle Ages*, Oxford and Cambridge: Blackwell.

Leblebici, Huseyin, Salancik, Gerard R., Copay, Anne and King, Tom (1991) Institutional change and the transformation of interorganizational fields: an organizational history of the U.S. radio broadcasting system, *Administrative Science Quarterly*, 36: 333–363.

Lefebvre, Henri (1991) *The Production of Space*, Oxford: Blackwell.

Leidner, Robin (1993) *Fast Food, Fast Talk: Service Work and the Routinization of Everyday Life*, Berkeley: University of California Press.

Leonard-Barton, D. (1995) *Wellspring of Knowledge: Building and Sustaining the Sources of Innovation*, Boston: Harvard Business School Press.

Levinson, H. (1996) Executive coaching, *Consulting Psychology Journal: Practice and Research*, 48(2): 115–123.

Lévi-Strauss, Claude (1992) *Tristes tropiques*, London: Penguin.

Lindberg, Kajsa and Czarniawska, Barbara (2006) Knotting the action net, organizing between organizations, *Scandinavian Journal of Management*, 22: 292–306.

Lindgren, Monika and Packendorff, Johann (2006) What's new in new forms of organizing? On the construction of gender in project-based work, *Journal of Management Studies*, 43(4): 841–866.

Lindkvist, Lars (2005) Knowledge communities and knowledge collectives. A typology of knowledge work in groups, *Journal of Management Studies*, 42(6): 1189–1210.

Lingard, Helen and Francis, Valerie (2004) The work-life experiences of office and site-based employees in the Australian construction industry, *Construction Management and Economics*, 22(9): 991–1002.

Lingard, Helen and Francis, Valerie (2006) Does a supportive work environment moderate the relationship between work–family conflict and burnout among construction professionals?, *Construction Management and Economics*, 24(2): 185–196.

Linstead, Stephen and Höpfl, Heather, Ed. (2000) *The Aesthetics of Organization*, London: Thousand Oaks and New Delhi: Sage.

Livingston, Eric (1986) *The Ethnomethodological Foundations of Mathematics*, London, Boston and Henley: Routledge & Kegan Paul.

Loosemore, Martin and Chau, D.W. (2002) Radical discrimination toward Asian operatives in the Australian construction industry, *Construction Management and Economics*, 20: 91–102.

Love, Peter E., Li, Heng, Irani, Zahir and Faniran, Olusegun (2000) Total quality management and the learning organization: a dialogue for change in construction, *Journal of Construction Research*, 18: 321–331.

Lowe, David and Skitmore, Martin (1994) Experiential learning in cost estimating, *Construction Management and Economics*, 12: 423–431.

Luhmann, Niklas (1979) *Trust and Power*, Wiley: New York.

Luhmann, Niklas (1982) *The Differentiation of Society*, New York: Columbia University Press.

Luhmann, Niklas (1990) *Essays on Self-Reference*, New York: Columbia University Press.

Luhmann, Niklas (1995) *Social Systems*, Stanford: Stanford University Press.

Luhmann, Niklas (2002) *Theories of Distinction: Redescribing the Descriptions of Modernity*, Stanford: Stanford University Press.

Luhmann, Niklas (2000a) *Art as a Social System*, trans. by Knodt, Eva M., Stanford: Stanford University Press.

Luhmann, Niklas (2000b) *The Reality of the Mass Media*, Cambridge: Polity Press.

Lundin, Rolf A. and Steinthórsson, Runólfur S. (2003) Studying organizations as temporary, *Scandinavian Journal of Management*, 19: 233–250.

Lynch, Michael (1985) Art and Artifact in Laboratory Science. *A Study of Shop Work and Shop Talk in a Research Laboratory*, London: Routledge & Kegan Paul.

Lynch, Michael (2002) Protocols, practices and the reproduction of technique in molecular biology, *British Journal of Sociology*, 53(2): 203–220.

Machlup, Fritz (1962) *The Production and Distribution of Knowledge in the United States*, Princeton: Princeton University Press.

MacIntyre, Alasdair (1981) *After Virtue*, London: Duckworth.

Mack, Kathy S. (2007) Senses of seascapes: aesthetics and the passion of knowledge, *Organization*, 14(3): 373–390.

Mackenzie, Adrian (2005) Problematizing the technological: the object as event?, *Social Epistemology*, 19(4): 381–399.

MacKenzie, Donald (1996) *Knowing Machines: Essays on Technical Change*, Cambridge, MA: MIT Press.

MacKenzie, Donald (1999) Slaying the Kraken: the sociohistory of a mathematical proof, *Social Studies of Science*, 29(1): 7–60.

Manning, Peter K. (1992) *Organizational communication*, New York: Aldine de Gruyter.

March, James G. and Olsen, Johan P. (1976) *Ambiguity and Choice in Organizations*, Oslo: Universitetsforlaget.

Marotto, Mark, Roos, Johan and Victor, Bart (2007) Collective virtuosity in organizations: a study of peak performance in an orchestra, *Journal of Management Studies*, 44(3): 388–413.

Maturana, Humberto R. and Varela, Francisco J. (1980) *Autopoesis and Cognition: The Realization of the Living*, Dordrecht, Boston and London: D. Riedle Publishing.

Maturana, Humberto R. and Varela, Francisco J. (1992) *The Tree of Knowledge: The Biological Roots of Human Understanding*, Boston and London: Shambala.

Maurer, Indre and Ebers, Mark (2007) Dynamics of social capital and their performance implications: lessons from biotechnology start-ups, *Administrative Science Quarterly*, 52: 262–292.

McDowell, Linda (1997) *Capital Culture: Gender at Work in the City*, Oxford and Malden: Blackwell.

McLuhan, Marshall (1962) *The Gutenberg Galaxy: The Making of Typographic Man*, London: Routledge & Kegan Paul.

Mead, George H. (1934) *Mind, Self, and Society*, Chicago: University of Chicago Press.

Merleau-Ponty, Maurice (1962) *Phenomenology of Perception*, London and New York: Routledge.

Merleau-Ponty, Maurice (1968) *The Visible and the Invisible*, Evanston: Northwestern University Press.

Merton, Robert K. (1973) *The Sociology of Science: Theoretical and Empirical Investigations*, Storer, Norman W., Ed. Chicago: The University of Chicago Press.

Meyer, John W. and Rowan, Brian (1977) Institutionalizing organizations: formal structure as myth and ceremony, *American Journal of Sociology*, 83(2): 340–363.

Michelson, Grant and Mouly, Suchitra (2000) Rumour and gossip in organisations: a conceptual study, *Management Decision*, 38(5): 339–346.

Miles, Matthew B. and Huberman, A. Michael (1994) Data management and analysis methods, in Denzin, Norman K. and Lincoln, Yvonne S. (1994) *Handbook of Qualitative Research*, London: Sage.

Mills, Charles Wright (1951) *White Collars: The American Middle Class*, Oxford: Oxford University Press.

Mills, Charles Wright (1956) *The Power Elite*, Oxford and New York: Oxford University Press.

Mills, Charles Wright (1959) *The Sociological Imagination*, Oxford: Oxford University Press.

Mintzberg, Henry (1973) *The Nature of Managerial Work*, New York: Harper and Row.

Mizruchi, Mark S. and Fein, Lisa C. (1999) The social construction of organizational knowledge: a study of the uses of coercive, mimetic, and normative isomorphism, *Administrative Science Quarterly*, 44: 653–683.

Mody, Cyrus C.M. (2005) The sounds of silence: listening to laboratory practice, *Science Technology and Human Values*, 30(2): 175–198.

Mol, Annemarie (2002) Cutting surgeons and walking patients. Some components involved in comparing, in Law, John and Mol, Annemarie, Eds. (2002) *Complexities: Social Studies of Knowledge Practices*, Durham and London: Duke University Press.

Mouw, Ted (2006) Estimating the social effect of social capital: a review of recent research, *Annual Review of Sociology*, 32: 79–102.

Mulvey, Laura (1989) Visual pleasures and narrative cinema, in Mulvey, Laura, Ed. (1989) *Visual and Other Pleasures*, London: Macmillan, pp. 14–26.

Murningham, J. Keith and Conlon, Donald E. (1991) The dynamics of intense work groups: a study of British String quartets, *Administrative Science Quarterly*, 36: 165–186.

Mustapha, F.H. and Naoum, A. (1998) Factors influencing the effectiveness of construction site managers, *International Journal of Project Management*, 16(1): 1–8.

Nahapiet, Janine and Ghosal, Sumantra (1998) Social capital, intellectual capital, and the organizational advantage, *Academy of Management Review*, 23(2): 242–266.

Nassehi, Armin (2005) Organizations as design machines: Niklas Luhmann's theory of organized social systems, in Jones, Campbell and Munro, Rolland, Eds. (2005) *Contemporary Organization Theory*, Oxford and Malden: Blackwell, pp. 178–191.

Nayak, Ali (2008) On the way to theory: a processual approach, *Organization Studies*, 29(2): 173–190.

Nelson, Richard R. and Winter, Sidney G. (1982) *An Evolutionary Theory of the Economic Change*, Cambridge, MA: Belknap.

Newell, Sue, Tansley, Carole and Huang, Jimmy (2004) Social capital and knowledge integration in an ERP project team: the importance of bridging and bonding, *British Journal of Management*, 15: S43–S57.

Nicolini, Davide (2007) Studying visual practices in construction, *Building Research and Information*, 35(5): 576–580.

Nietzsche, Friedrich (1886/1990) *Beyond Good and Evil*, Penguin, London.

Nixon, Sean (2005) *Advertisement Culture*, London: Thousand Oaks and New Delhi: Sage.

Nonaka, Ikujiro and Takeuchi, Hirotaka (1995) *The Knowledge-Creating Company*, Oxford: Oxford University Press.

Nooteboom, Bart (2000) Institutions and forms of co-ordination in innovative systems, *Organization Studies*, 21(5): 915–939.

Obrist, Hans Ulrich and Koolhaas, Rem (2001) Relearning from Las Vegas: an interview with Denise Scott Brown and Robert Venturi, in Chung, Chuihua Judy, Inaba, Jeffrey, Koolhaas, Rem and Leong, Sze Tsung, Eds. (2001) *Harvard Design School Guide to Shopping*, Köln: Taschen.

Ofori, George and Kien, Ho Lay (2004). Translating Singapore architects' environmental awareness into decision making, *Building Research and Information*, 32(1): 27–37.

Ogulana, Olu (1991) Learning from experience in designing cost estimating, *Construction Management and Economics*, 9: 133–150.

Ong, Walter J. (1982) *Orality and Literacy: The Technologizing of the Word*, London: Routledge.

Orlikowski, Wanda J. (2002) Knowing in practice. Enacting a collective capability in distributed organizing, *Organization Science*, 13(3): 249–273.

Orlikowski, Wanda J. (2007) Sociomaterial practices: exploring technology at work, *Organization Studies*, 28(9): 1435–1448.

Orr, Julian E. (1996) *Talking about Machines: An Ethnography of a Modern Job*, Ithaca and London: Cornell University Press.

O'Shaughnessy, S. (2001) Executive coaching: the route to business stardom, *Industrial and Commercial Training*, 33(6): 194–197.

Owen-Smith, Jason (2001) Managing laboratory work through scepticism: processes of evaluation and control, *American Sociological Review*, 66: 427–452.

Palmer, B. (2003) Maximizing value from executive coaching, *Strategic HR Review*, 2(6): 26–29.

Parsons, Talcott (1934/1990) Prolegomena to a theory of social institutions, *American Sociological Review* 55: 319–33.

Parsons, Talcott (1951/1991) *The Social System*, London and New York: Routledge.

Patriotta, Gerardo (2003) Sensemaking on the shop floor: narratives of knowledge in organizations, *Journal of Management Studies*, 40(2): 349–375.

Paules, Greta Foff (1991) *Dishing It Out; Power and Resistance Among Waitresses in a New Jersey Restaurant*, Philadelphia: Temple University Press.

Pels, Dick, Hetherington, Kevin and Vendenberghe, Frédéric (2002) The status of the object: performances, mediations and techniques, *Theory, Culture and Society*, 19(5/6): 1–21.

Pentland, Brian T. and Rueter, Henry H. (1994) Organization routines as grammars of action, *Administrative Science Quarterly*, 39(3): 484–510.

Perlow, Leslie A. (1999) The time famine; toward a sociology of work time, *Administrative Science Quarterly*, 44: 57–81.

Peterson, D.B. (1996) Executive coaching at work: the art of one-on-one change, *Consulting Psychology Journal: Practice and Research*, 48(2): 78–86.

Pettigrew, Andrew M. (1973) *The Politics of Organizational Decision-Making*, London: Tavistock.

Pfeffer, Jeffrey and Salancik, Gerard R. (1978) *The External Control of Organizations: A Resource Dependence Perspective*, New York: Harper and Row.

Pickering, Andrew (1995) *The Mangle of Practice: Time, Agency, and Science*, Chicago and London: The University of Chicago Press.

Piñeiro, Erik (2007) Aesthetics at the heart of logic: on the role of beauty in computing innovation, Guillet de Monthoux, Pierre, Gustafsson, Claes and Sjöstrand, Sven-Erik, Eds. (2007) *Aesthetic Leadership: Managing Fields of Flow in Art and Business*, Basingstoke: Palgrave Macmillan, pp. 105–127.

Pinnington, Ashly and Morris, Timothy (2002) Transforming the architect. Ownership from the archetype change, *Organization Studies*, 23(2): 189–210.

Porcello, Thomas (2004) Speaking of sound: language and the professionalization of sound-recording engineers, *Social Studies of Science*, 34(5): 733–758.

Portes, Alejandro (1998) Social capital: its origin and applications in modern sociology, *Annual Review of Sociology*, 23: 1–27.

Powell, Walter W. (1985) *Getting into Print: The Decision Process in Scholarly Publishing*, Chicago and London: The University of Chicago Press.

Powell, Walter W. and Snellman, Kaisa (2004) The knowledge economy, *Annual Review of Sociology*, 30: 199–220.

Prasad, Amit (2005) Making images/making bodies: visibility and disciplining through magnetic resonance imaging (MRI), *Science, Technology and Human Values*, 30(2): 291–316.

Prasad, Pushkala (2005) *Crafting Qualitative Research: Working in the Postpositivist Traditions*, New York: Armonk and London: ME Sharpe.

Prentice, Rachel (2005) The anatomy of surgical simulations: the mutual articulation of bodies in and through the machine, *Social Studies of Science*, 35(6): 837–866.

Rabinow, Paul (1996) *Making PCR: A Story of Biotechnology*, Chicago and London: The University of Chicago Press.

Radcliffe-Brown, Alfred R. (1958) *Methods in Social Anthropology*, Chicago: The University of Chicago Press.

Räisänen, Christine and Linde, Anneli (2004) Technologizing discourse to standardize projects in multi-project organizations: hegemony by consensus, *Organization*, 11(1): 101–121.

Reason, Peter (2006) Choice and quality in action research practice, *Journal of Management Inquiry*, 15(2): 187–203.

Reich, Robert B. (1991) *The Work of Nations*, London: Simon & Schuster.

Rheinberger, Hans-Jörg (1997) *Toward a History of Epistemic Things: Synthesizing Proteins in the Test Tube*, Stanford: Stanford University Press.

Robertson, Maxine and Swan, Jacky (2003) 'Control – what control?' Culture and ambiguity within a knowledge intensive firm, *Journal of Management Studies*, 40(4): 831–858.

Robertson, Maxine and Swan, Jacky (2004) Going public: the emergence and effects of soft bureaucracy within a knowledge-intensive firm, *Organization*, 11(1): 123–148.

Robinson, Herbert S., Carrillo, Patricia M., Anumba, Chimay P., Al-Ghassani, Ahmed M. (2005) Knowledge management in large construction companies, *Engineering, Construction, and Architectural Management,* 12(5): 431–445.

Rooke, John and Clark, Leslie (2005) Learning, knowledge and authority on site: a case of safety practice, *Building Research and Information,* 33(6): 561–570.

Rooke, John, Seymour, David and Fellows, Richard (2004) Planning for claims: an ethnography of industry culture, *Construction Management and Economics,* 22: 655–662.

Rorty, Richard (1998) *Truth and Progress: Philosophical papers,* Vol. 3, Cambridge: Cambridge University Press.

Roth, Jonas (2003) Enabling knowledge creation: learning from an R&D organization, *Journal of Knowledge Management,* 7(1): 32–48.

Roy, Donald (1952) Quote restriction and goldbricking in a machine shop, *American Journal of Sociology,* 57(5): 427–442.

Ryle, G. (1949) *The Concept of Mind,* Harmondsworth: Penguin.

Sackmann, Sonja A. (1992) Culture and subcultures: an analysis of organizational knowledge, *Administrative Science Quarterly,* 37: 140–161.

Saint, Andrew (1983) *The Image of the Architect,* New Haven: Yale University Press.

Salzer-Mörling, Miriam (2002) Changing corporate landscapes, in Holmberg, Ingagill, Salzer-Mörling, Miriam and Strannegård, Lars, Eds. (2002) *Stuck in the Future: Tracing the 'New Economy',* Stockholm: Bookhouse Publishing.

Sanders, Teela (2004) Controllable laughter: managing sex work through humour, *Sociology,* 38(2): 273–291.

Sang, Katherine J.C., Dainty, Andrew J., Ison, Stephen G. (2007) Gender: a risk factor for occupational stress in the architectural profession?, *Construction Management and Economics,* 25: 1305–1317.

Scarbrough, Harry and Swan, Jacky (2001) Explaining the diffusion of kowledge management: the role of fashion, *British Journal of Management,* 12: 3–12.

Schatzki, Theodore R. (2002) *The Site of the Social: A Philosophical Account of the Constitution of Social Life and Change,* University Park, PA: The Pennsylvania State University Press.

Schatzki, Theodore R., Knorr Cetina, Karin and Savigny, Eike von, Eds. (2001) *The Practice Turn in Contemporary Theory,* London and New York: Routledge.

Schiller, Friedrich (1795/2004) *On the Aesthetic Education of Man,* Mineola: Dover Publications.

Schleef, Debra J. (2006) *Managing Elites: Professional Socialization in Law and Business Schools,* Lanham: Rowman and Littlefield.

Schultze, Ulrike (2000) A confessional account of an ethnography about knowledge work, *MIS Quarterly,* 24(1): 3–41.

Schütz, Alfred (1962) *Collected Papers,* Vol. I, *The Problem of Social Reality,* The Hague: Martinus Nijhoff.

Scott, Richard W. (2004) Reflections on a half-century of organizational sociology, *Annual Review of Sociology,* 30: 1–21.

Scott, W. Richard (1995) *Institutions and Organizations,* Thousand Oaks, London and New Delhi: Sage.

Scott, W. Richard (2008) Lords of the dance: professionals as institutional agents, *Organization Studies,* 29(2): 219–238.

Selznick, Philip (1949) *TVA and the Grassroots,* Berkeley: University of California Press.

Selznick, Philip (1957) *Leadership in Administration*, Berkeley: University of California Press.

Selznick, Philip (1996) Institutionalism 'old' and 'new', *Administrative Science Quarterly*, 40: 270–277.

Sennett, Richard (1998) *The Corrosion of Character: The Personal Consequences of Work in the New Capitalism*, New York and London: W.W. Norton and Company.

Serres, Michel (1991) *Rome: The Book of Foundations*, trans. by McCarren, Felicia, Stanford: Stanford University Press.

Shapin, Steven (1994) *A Social History of Truth: Civility and Science in Seventeenth-century England*, Chicago and London: University of Chicago Press.

Shenhav, Yahouda (1995) From chaos to systems: the engineering foundations of organization theory, 1879–1932, *Administrative Science Quarterly*, 40: 447–585.

Shenhav, Yehouda (1999) Manufacturing Rationality. *The Engineering Foundation of the Managerial Revolution*, Oxford and New York: Oxford University Press.

Shusterman, Richard (2006) *Aesthetics, Theory, Culture and Society*, 23(2–3): 237–252.

Siedl, David (2005) The basic concepts of Luhmann's theory of social systems, in Seidl, David and Becker, Kai Helge, Eds. (2005) *Niklas Luhmann and Organisation Studies*, Copenhagen: Copenhagen Business School Press, pp. 21–53.

Silverman, David (1993) *Interpreting Qualitative Data*, London: Sage.

Simon, Herbert A. (1991) Bounded rationality and organizational learning, *Organization Science*, 2(1): 125–133.

Smith, Ryan A. (2002) Race, gender and authority in the workplace. Theory and research. *Annual Review of Sociology*, 28(2): 509–542.

Söderlund, Jonas (2004) Building theories of project management: past research, questions for the future, *International Journal of Project Management*, 22: 183–191.

Sole, Deborah and Edmondson, Amy (2002) Situated knowledge and learning in disperse teams, *British Journal of Management*, 13: S17–S34.

Spender, J.C. and Grant, Robert M. (1996) Knowledge and the firm: overview, *Strategic Management Journal*, 17(Winter Special Issue): 5–9.

Stake, Robert E. (1996) *The Art of Case Study Research*, Thousand Oaks, London and New Delhi: Sage.

Star, Susan Leigh and Griesemer, James R. (1989) Institutional ecology, 'translation' and boundary objects: amateurs and professionals in Berlely's Museum of Vertebrate Zoology, 1907–1039, *Social Studies of Science*, 19(3): 387–420.

Starbuck, William H. (1992) Learning by knowledge-intensive firms, *Journal of Management Studies*, 29(6): 713–740.

Starkey, Ken and Madan, Paula (2001) Bridging the relevance gap: aligning stake-holders in the future of management research, *British Journal of Management*, 12(Special Issue): S3–S26.

Steyrer, Johannes (1998) Charisma and the archetypes of leadership, *Organization Studies*, 19(5): 807–828.

Stinchcombe, Arthur (1959) Bureaucractic and craft administration of production. A comparative study, *Administrative Science Quarterly*, 4(2): 168–188.

Strati, Antonio (1999) *Organization and Aesthetics*, London, Thousand Oaks and New Delhi: Sage.

Strauss, Anselm L. and Corbin, Juliet (1998) *Basics of Qualitative Research*, 2nd edn, London, Thousand Oaks and New Delhi: Sage.

Strauss, Anselm, Schatzman, Leonard, Bucher, Rue, Ehrlich, Danuta and Sabshin, Melvin (1964) *Psychiatric Ideologies and Institutions*, 2nd edn, New Brunswick and London: Transaction Books.

Styhre, Alexander (2003) *Understanding Knowledge Management: Critical and Postmodern Perspectives*, Copenhagen: Copenhagen Business School Press.

Styhre, Alexander (2006) The bureaucratization of the project manager function: the case of construction industry, *International Journal Project Management*, 24: 271–276.

Styhre, Alexander and Josephson, Per-Erik (2006) Revisiting site manager work: stuck in the middle?, *Construction Management and Economics*, 24: 521–528.

Styhre, Alexander and Sundgren, Mats (2005) *Managing Creativity in Organizations: Critique and Practices*, Basingstoke: Palgrave.

Styhre, Alexander, Roth, Jonas and Ingelgård, Anders (2002) Care of the other: knowledge creation through care in professional teams, *Scandinavian Journal of Management*, 18(4): 503–520.

Subramaniam, Mohan and Youndt, Mark A. (2005) The influence of intellectual capital on the types of innovative capabilities, *Academy of Management Journal*, 48(3): 450–463.

Sudnow, David (1978) *Ways of the Hand: The Organization of Improvised Conduct*, Cambridge, MA: Harvard University Press.

Sundgren, Mats and Styhre, Alexander (2006) Leadership as de-paradoxification: Leading new drug development work at three pharmaceutical companies, *Leadership*, 2(1): 31–52.

Sverlinger, Per-Olof M. (2000) *Managing Knowledge in Professional Service Organizations: Technical Consultants Serving the Construction Industry*, Ph.D. thesis, Department of Service Management, Chalmers University of Technology.

Symes, Martin, Eley, Joanna and Seidel, Andrew D. (1995) *Architects and their Practice*, London: Butterworth.

Szulanski, Gabriel (1996) Exploring internal stickiness: impediments to the transfer of best practice within the firm, *Strategic Management Journal*, 17(Winter Special Issue): 27–43.

Szulanski, Gabriel and Cappetta, Rosella (2003) Stickiness: conceptualizing, measuring, and predicting difficulties in the transfer of knowledge within organizations, in Easterby-Smith, Mark and Lyles, Marjorie A, *Handbook of Organization Learning and Knowledge Management*, Oxford and Malden: Blackwell, pp. 513–534.

Taylor, Steven S. and Hansen, Hans (2005) Finding form: looking at the field of organizational aesthetics, *Journal of Management Studies*, 42(6): 1211–1231.

Teece, David J. (2000) *Managing Intellectual Capital: Organizational, Strategic and Policy Dimensions*, Oxford and New York: Oxford University Press.

Tengblad, Stefan (2002) Time and space in managerial work, *Scandinavian Journal of Management*, 18: 543–565.

Terkel, Studs (1972) *Working*, New York: The New Press.

Thomas, Alan B. (2004) The coming crisis of Western management education, in Jeffcut, Paul, Ed. (2004) *The Foundations of Management Knowledge*, London and New York: Routledge.

Thomas, Robin and Linstead, Alison (2002) Losing the plot? Middle management and identity, *Organization*, 9(1): 71–93.

Timmermans, Stefan and Berg, Marc (1997) Standardization in action: achieving local universality through medical protocols, *Social Studies of Science*, 27(2): 273–305.

Tolbert, Pamela S. and Zucker, Lynne G. (1996) The institutionalization of institutional theory, in Clegg, Stewart R., Hardy, Cynthia and Nord, Walter R., Eds. (1996) *Handbook of Organizational Studies*, London, Thousand Oaks and New Delhi: Sage.

Touraine, Alain (1971) *The Post-Industrial Society. Tomorrow's Social History: Classes, Conflicts and Culture in the Programmed Society*, trans. by Mayhew, Leonard F.X., New York: Random House.

Traweek, Sharon (1988) *Beamtimes and Lifetimes: The World of High Energy Physicists*, Cambridge, MA and London: Harvard University Press.

Trieb, Marc (1996) *Space Calculated in Seconds: The Philips Pavilion, Le Corbusier, Edgard Varèse*, Princeton: Princeton University Press.

Tsoukas, Haridimos (1996) The firm as distributed knowledge system: a constructionist approach, *Strategic Management Journal*, 17(Winter Special Issue): 11–25.

Tsoukas, Haridimos (2005) *Complex Knowledge: Studies in Organizational Epistemology*, Oxford and New York: Oxford University Press.

Tsoukas, Haridimos and Mylonopoulos, Nikolaus (2004) Introduction: What does it mean to view organizations as knowledge systems?, in Tsoukas, Haridimos and Mylonopoulos, Nikolaus, Eds. (2004) *Organization Knowledge Systems: Knowledge, Learning and Dynamic Capabilities*, Basingstoke and New York: Palgrave, pp. 1–26.

Tsoukas, Haridimous and Mylonopoloulos, Nikos (2004) Introduction: knowledge construction and creation in organizations, *British Journal of Management*, 15: S1–S8.

Tsoukas, Haridimos and Vladimirou, Efi (2001) What is organizational knowledge?, *Journal of Management Studies*, 38(7): 973–993.

Tyler, Melissa and Taylor, Steve (1998) The exchange of aesthetics: women's work and 'the gift', *Gender, Work and Organization*, 5(3): 165–171.

Ullman, Ellen (1997) *Close to the Machine: Technophilia and its Discontents*, San Francisco: City Lights Book.

United States Office of Personnel Management (1998) *Handbook of Occupational Groups and Families*, URL: http://www.opm.gov/fedclass/text/HdBkToC.htm (accessed October 11, 2007).

Unwin, Simon (2007) Analysing architecture through drawing, *Building Research and Information*, 35(1): 101–110.

Van Delinder, Jean (2005) Taylorism, managerial control strategies, and the ballets of Balanchine and Stravinsky, *American Behavioral Scientist*, 48(11): 1439–1452.

Van Maanen, John (1975) Police socialization: a longitudinal examination of job attitudes in an urban police department, *Administrative Science Quarterly*, 20: 207–228.

Vann, Katie and Bowker, Geoffrey C. (2001) Instrumentalizing the truth of practice, *Social Epistemology*, 15(3): 247–262.

Venturi, Robert, Brown, Denise Scott and Izenour, Steven (1977) *Learning from Las Vegas: The Forgotten Symbolism of Architectural Form*, Cambridge, MA: MIT Press.

Von Hippel, Eric (1998) Economics of product development by users: the impact of 'sticky' local information, *Management Science*, 44(5): 629–644.

von Krogh, G. (1998) Care in knowledge creation, *California Management Review*, 40(3): 133–153.

Warren, Samatha (2008) Empirical challenges in organizational aesthetics; towards a sensual methodology, *Organization Studies*, 29(4): 559–580.

Wasylyshyn, Karol M. (2003) Executive coaching: an outcome study, *Consulting Psychology Journal: Practice and Research*, 55(2): 94–106.

Weber, Max (1948) Science as a vocation, in Gerth, H.H. and Mills, Charles Wright, Eds. (1948) *From Max Weber: Essays in Sociology*, London: Routledge & Kegan Paul, pp. 129–156.

Weick, Karl E. and Roberts, Karlene H. (1993) Collective mind in organizations: heedful interrelating on flight decks, *Administrative Science Quarterly*, 38: 357–381.

Weick, Karl E. (1976) Educational organizations as loosely coupled systems, *Administrative Science Quarterly,* 21: 1–19.

Weick, Karl E. (1979) *The Social Psychology of Organizing*, 2nd edn, New York: McGraw-Hill.

Weick, Karl E. and Sutcliffe, Kathleen M. (2006) Mindfulness and the quality of organizational attention, *Organization Science*, 17(4): 514–524.

Wenger, Etienne (2000) Communities of practice and social learning systems, *Organization*, 7(2): 225–246.

Wenger, Etienne, McDermott, Richard and Snyder, William M. (2002) *Cultivating Communities of Practice*, Boston: Harvard Business School Press.

White, Hayden (1987) *The Content of Form: Narrative Discourse and Historical Representation*, Baltimore and London: John Hopkins University Press.

Whitehead, Alfred North (1967) *Adventures of Ideas,* New York: Free Press.

Whitley, Richard (2000) The institutional structuring of innovation strategies: business systems, firm types, and patterns of technical change in different market economies, *Organization Studies*, 21(5): 855–886.

Whyte, Jennifer K., Ewenstein, Boris, Hales, Michael and Tidd, Joe (2007) Visual practices and the objects used in design, *Building Research and Information*, 35(1): 18–27.

Whyte, William H. (1956) *The Organization Man*, Simon & Schuster, New York.

Wilcox King, Adelaide, Fowler, Sally W. and Zeithaml, Carl P. (2001) Managing organizational competencies for competitive advantage: the middle-management edge, *Academy of Management Executive*, 15(2): 95–106.

Wilemon, David L. and Cicero, John P. (1970) The project manager: anomalies and ambiguities, *Academy of Management Journal*, 13: 269–282.

Willem, Annick and Scarbough, Harry (2006) Social capital and political bias in knowledge sharing: an exploratory study, *Human Relations*, 59(10): 1343–1370.

Willis, Paul (1977) *Learning to Labour*, Farnborough, Saxon House.

Winch, Graham (1998) Zephyrs of creative destruction: understanding the management of innovation in construction, *Building Research and Information*, 26(4): 268–297.

Winch, Graham M. (2003) How innovative is construction? Comparing aggregated data on construction innovation and other sectors – A case of apples and pears. *Construction Management and Economics*, 21: 651–654.

Winch, Graham and Schneider, Eric (1993) Managing the knowledge-based organization: the case of architectural practice, *Journal of Management Studies*, 30(6): 923–937.

Yaneva, Albena (2005) Scaling up and down: extraction trails in architectural design, *Social Studies of Science*, 35(6): 867–894.

Yanow, Dvora (2004) Translating local knowledge at organizational peripheries, *British Journal of Management*, 15: S9–S25.

Yli-Renka, Helena, Autio, Erkko and Sapienza, Harry J. (2001) Social capital, knowledge acquisition, and knowledge exploitation in young technology-based firms, *Strategic Management Journal*, 22: 587–613.

Žižek, Slavoj (1995) *The Metastases of Enjoyment*, London: Verso.

Zucker, Lynne G. (1987) Institutional theories of organizations, *Annual Review of Sociology*, 13: 443–464.

Index

eBooks – at www.eBookstore.tandf.co.uk

A library at your fingertips!

eBooks are electronic versions of printed books. You can store them on your PC/laptop or browse them online.

They have advantages for anyone needing rapid access to a wide variety of published, copyright information.

eBooks can help your research by enabling you to bookmark chapters, annotate text and use instant searches to find specific words or phrases. Several eBook files would fit on even a small laptop or PDA.

NEW: Save money by eSubscribing: cheap, online access to any eBook for as long as you need it.

Annual subscription packages

We now offer special low-cost bulk subscriptions to packages of eBooks in certain subject areas. These are available to libraries or to individuals.

For more information please contact webmaster.ebooks@tandf.co.uk

We're continually developing the eBook concept, so keep up to date by visiting the website.

www.eBookstore.tandf.co.uk